T0205495

Energy Recovery Processes from Wastes

Sadhan Kumar Ghosh
Editor

Energy Recovery Processes from Wastes

 Springer

Editor
Sadhan Kumar Ghosh
Department of Mechanical Engineering
Jadavpur University
Kolkata, West Bengal, India

ISBN 978-981-32-9230-7   ISBN 978-981-32-9228-4   (eBook)
https://doi.org/10.1007/978-981-32-9228-4

This Springer imprint is published by the registered company Springer Nature Singapore Pte Ltd.
The registered company address is: 152 Beach Road, #21-01/04 Gateway East, Singapore 189721, Singapore

Preface/Editorial: Energy Recovery Processes from Wastes

The current global municipal solid waste (MSW) generation levels are approximately 1.3 billion tons per year and are expected to increase to approximately 2.2 billion tons per year by 2025. Thus, with the growth of society, it is becoming an inevitable by-product, and one of the greatest challenges for future generations is to manage these large quantities of waste in a sustainable way. One approach is to minimise the amount of waste produced and to recycle larger fractions of waste materials. A considerable part of the waste is left out and needs a better solution than landfilling. The problem of waste management sector cannot be solved by itself; on the other hand, the energy sector is a perfect answer with ever-increasing energy demands. Energy recovery technologies can provide alternative energy, reduce the burden on landfill site and mitigate greenhouse gas emissions. A variety of processes exist for waste conversions. The most used of these are thermal conversions {incineration, fast and slow pyrolysis, gasification, production of refuse-derived fuel (RDF), in the forms of methane, hydrogen and other synthetic fuels (biochemical, e.g. anaerobic digestion/biomethanation, mechanical biological treatment—composting, vermicomposting, refuse-derived fuel, etc.)} and chemical conversions (trans-esterification and other processes to convert plant and vegetable oils to biodiesel). The choice of conversion process depends on the type, property and quantity of biomass feedstock, the desired form of the energy, end-use requirements, environmental standards, economic conditions and project-specific factors.

Waste can act as a major source of energy rather than a disposable material. Segregated and processed waste in the form of RDF/SRF can act as an effective alternative fuel, having comparable energy potential to coal. A sustainable waste management strategy should extract the full energy and environmental value from MSW. As a result—and despite a recent economic crisis—the global market for WtE technologies has experienced substantial growth. Waste-to-energy plants are not merely incinerators: there, the waste is recycled as energy in the form of electricity and heat. The consumption of primary resources is consequently reduced. A worldwide growth market is strategically expanding. The market for thermal treatment and energetic recovery of residual and other types of solid waste is growing day by day. Increasing waste amounts, shrinking landfill spaces in agglomerations and higher ecological standards stimulate this growth throughout the world. Presently, more than 2430 WtE plants are active worldwide in more than 40 countries having a disposal capacity of around 360 million tons of waste per year. Only in 2017, more than 80 new thermal treatment plants have been installed with a total treatment capacity of 25 million tons per year. More than 2700 plants have been estimated with a capacity of about 530 million tons per year to be operational by 2027. In USA only, 86 WtE plants have the capacity to produce 2720 megawatts of power per year by processing more than 28 million tons. Technological advancement, improved pollution control systems, governmental incentives and stringent regulations have made WtE technology a potential alternative, especially for the developed countries. If 1 ton of MSW is incinerated for electricity generation instead of landfilling (without gas recovery), then 1.3 tons of CO_2 equivalent emissions can be avoided if equivalent CO_2 emissions from fossil-fuel-based power plants are also considered to generate the same amount of electricity (ASME, 2008).

The 8th IconSWM 2018 has received 380 abstracts and 320 full papers from 30 countries. Three hundred accepted full papers have been presented in oral and poster presentations in November 2018 in Guntur, Andhra Pradesh, India. After the thorough review by experts and required revisions, the board has finally selected 20 chapters in this book, *Energy Recovery Processes from Wastes* dealing with the design of reactor and solar thermal steam generator, gas fermenter, plasma arc gasification, thermolysis of waste polystyrene, waste pyrolysis, application of waste HDPE oil and diesel–biodiesel blends, modelling and simulation of biodiesel, thermochemical conversion of different feedstocks, production of biogas and bioethanol, waste-to-energy conversion technologies, co-processing of wastes in cement plants and economic feasibility of bioenergy systems.

The IconSWM movement was initiated for better waste management, resource circulation and environmental protection since the year 2009 through generating awareness and bringing all the stakeholders together from all over the world under the aegis of the International Society of Waste Management, Air and Water (ISWMAW). It established a few research projects across the country those include the CST at Indian Institute of Science, Jadavpur University, etc. Consortium of Researchers in International Collaboration (CRIC) and many other organisations across the world helped in the IconSWM movement. IconSWM has become significantly one of the biggest platforms in India for knowledge sharing, awareness generation and encouraging the urban local bodies (ULBs), government departments, researchers, industries, NGOs, communities and other stakeholders in the area of waste management. The primary agenda of this conference is to reduce the waste generation encouraging the implementation of 3Rs (Reduce, Reuse and Recycle) concept and management of the generated waste ensuring resource circulation. The conference has shown a paradigm and provided holistic pathways to waste management and resource circulation conforming to urban mining and circular economy.

The success of the 8th IconSWM is the result of significant contribution of many organisations and individuals, specifically the Government of Andhra Pradesh, several industry associations, chamber of commerce and industries, AP State Council of Higher Education and various organisations in India and in different countries as our partners including the UNEP, UNIDO and UNCRD. The 8th IconSWM 2018 was attended by nearly 823 delegates from 22 countries. The 9th IconSWM 2019 will be held at KIIT, Bhubaneswar, Odisha, during 27–30 November 2019, expecting nearly 900 participants from 30 countries.

This book will be helpful for the researchers, educational and research institutes, policy-makers, government, implementers, ULBs and NGOs.

Hope to see you all in 9th IconSWM 2019 in November 2019.

Kolkata, India   Prof. Sadhan Kumar Ghosh
June 2019

Acknowledgements

I gratefully acknowledge Shri Chandra Babu Naidu, Honourable Chief Minister, and Dr. P. Narayana, Honourable Minister of MA&UD, for taking personal interest in this conference.

I am indebted to Shri. R. Valavan Karikal, IAS, Dr. C. L. Venkata Rao, Shri. B. S. S. Prasad, IFS (Retd.), Prof. S. Vijaya Raju and Prof. A. Rajendra Prasad, VC, ANU, for their unconditional support and guidance for preparing the platform for successful 8th IconSWM at Guntur, Vijayawada, AP.

I must express my gratitude to Mr. Vinod Kumar Jindal, ICoAS; Shri. D. Muralidhar Reddy, IAS; Shri, K. Kanna Babu, IAS; Mr. Vivek Jadav, IAS; Mr. Anjum Parwez, IAS; Prof. S. Varadarajan, Mr. Bala Kishore and Mr. K. Vinayakam, Prof. Shinichi Sakai, Kyoto University, JSMCWM; Prof. Y. C. Seo and Prof. S. W. Rhee, KSWM; Shri. C. R. C. Mohanty, UNCRD; Members of Industry Associations in Andhra Pradesh; Prof. P. Agamuthu, WM&R; Prof. M. Nelles, Rostock University; Dr. Rene Van Berkel of UNIDO; and Ms. Kakuko Nagatani-Yoshida and Mr. Atul Bagai of UNEP and UN Delegation to India for their active support.

IconSWM-ISWMAW Committee acknowledges the contribution and interest of all the sponsors; industry partners; industries; co-organisers; organising partners around the world; Government of Andhra Pradesh; Swachha Andhra Corporation as the principal collaborator; Vice Chancellor and all the professors and academic community at Acharya Nagarjuna University (ANU); the Chairman, Vice Chairman, Secretary and other officers of AP State Council of Higher Education for involving all the universities in the state; the Chairman, Member Secretary and the officers of the AP Pollution Control Board; the director of factories; the director of boilers; director of mines and officers of different ports in Andhra Pradesh; and the delegates and service providers for making a successful 8th IconSWM.

I must specially mention the support and guidance by each of the members of the international scientific committee, CRIC members, the core group members and the local organising committee members of the 8th IconSWM who are the pillars for the success of the programme. The editorial board members including the reviewers, authors and speakers and Mr. Aninda Bose and Ms. Kamiya Khatter of M/s. Springer India Pvt. Ltd. deserve thanks who were very enthusiastic in giving me inputs to bring this book.

I must mention the active participation of all the team members in IconSWM Secretariat across the country with special mention of Prof. H. N. Chanakya and his team in IISc Bangalore; Ms. Sheetal Singh and Dr. Sandhya Jaykumar and their team in CMAK & BBMP; and Mr. Saikesh Paruchuri, Mr. Anjaneyulu, Ms. Senophiah Mary, Mr. Rahul Baidya, Ms. Ipsita Saha, Mr. Suresh Mondal, Mr. Bisweswar Ghosh, Mr. Gobinda Debnath and the research team members in Mechanical Engineering Department and ISWMAW, Kolkata HQ, for various activities for the success of the 8th IconSWM 2018.

I express my special thanks to Sannidhya Kumar Ghosh, being the governing body member of ISWMAW, for supporting the activities from USA. I am indebted to Mrs. Pranati Ghosh who gave me guidance and moral support in achieving the success of the event. Once again, the IconSWM and ISWMAW express gratitude to all the stakeholders, delegates and speakers who were the part of the success of the 8th IconSWM 2018.

Contents

About the Editor

Dr. Sadhan Kumar Ghosh is a Professor & Former Head of the Mechanical Engineering Department and the Founder Coordinator of the Centre for QMS at Jadavpur University, India. He is a prominent figure in the fields of waste management, circular economy, SME sustainability, green manufacturing, green factories, TQM and ISO standards. He served as the Director, CBWE, Ministry of Labour and Employment, Government of India and L&T Ltd. Prof Ghosh is the Founder and Chairman of the IconSWM and the President of the International Society of Waste Management, Air and Water, as well the Chairman of the "Indian Congress on Quality, Environment, Energy and Safety Management Systems (ICQESMS)" and the "Consortium of Researchers in International Collaboration" (CRIC). In 2012, he was awarded a Distinguished Visiting Fellowship by the Royal Academy of Engineering, UK, to work on "Energy Recovery from Municipal Solid Waste". He received the Boston Pledge and NABC 2006 award for the most eco-friendly innovation "Conversion of plastics & jute waste to wealth" in Houston, USA. He received patents on a waste plastic processing technology and a high-speed jute ribboning technology for preventing water wastage & occupational health hazards. He has published more than 50 research papers in leading international refereed journals and edited more than 30 books and proceedings. He is the Associate Editor of Journal of Japan Society of Materials Cycles and Waste Management. He has accomplished and has been involved in several impactful interdisciplinary research projects on sustainable supply chain of small- and medium-sized enterprises (SMEs) across the globe, circular economy in 34 countries, waste to energy and waste management. His projects have been funded by British Council, Royal Society, Royal Academy of Engineering, EU Horizon 2020, Jute Technology Mission, Central Pollution Control Board, Government of India, UNCRD/DESA, APO and Shota Rustaveli National Science Foundation (SRNSF) of Georgia.

Three-Stage Reactor Design to Convert MSW to Methanol

Aastha Paliwal and Hoysall N. Chanakya

Abstract Implementation of decentralized waste management requires quick and energy positive modes of waste reduction. Syngas conversion provides quick volume reduction and energy-rich gas. However, lack of infrastructure for the use of syngas at small scale makes this route less attractive. Conversion of syngas to locally usable methane and methanol, enabling on source application of energy, is an easy solution. We propose a three-stage reactor for the conversion of MSW, surplus agro-residue to methanol. Gasification, biomethanation, and methane fermentation to methanol represent the three stages of the process. S&T for the conversion of MSW, surplus agro-residue to syngas are in place. Therefore, this paper discusses the remaining two stages of the reactor, i.e., biomethanation via solid-state fermenter and methane fermentation to methanol via hanging rope reactor and its potential in decentralized production of a drop-in fuel, methanol from MSW rejects.

Keywords Methanol · Bioreactor · Syngas to methanol · Biofilm reactor · Gas to fuel · Solid-state reactor

1 Introduction

Population of 1.3 billion increasing at an annual rate of 1.3% [1] and 62 million tons of MSW per annum increasing at annual rate of 4% suggest decentralized waste treatment is the most sensible option to deal with quantum of waste at hand [2]. However, the absence of rapid waste reduction and energy positive process at small scale make 'decentralized' waste treatment an unattractive option at present. Gasification (to syngas) and biomethanation are two well-established waste-to-energy technologies, but they come at a price. While gasification is a faster way of converting waste to energy and has broader feedstock, the end product, syngas, is locally unusable; biomethanation provides ready-to-use fuel—biogas—but at long SRT of

A. Paliwal · H. N. Chanakya (✉)
Centre for Sustainable Technologies, Indian Institute of Science, Bangalore, India
e-mail: chanakya@iisc.ac.in

A. Paliwal
e-mail: aasthap@iisc.ac.in

© Springer Nature Singapore Pte Ltd. 2020
S. K. Ghosh (ed.), *Energy Recovery Processes from Wastes*,
https://doi.org/10.1007/978-981-32-9228-4_1

more than 30 days [3]. In India, wet MSW is best converted to biogas, while non-recyclable dry wastes are best converted to syngas, providing a solution for over 95% of our MSW. Converting syngas from gasifier to a locally usable, drop in fuel like methane/methanol at small scales [10–200 tpd] is a better way forward. Chemical conversion routes have high operating cost (200–900 °C, 5–20 MPa, expensive catalyst) [4] and hence are not feasible at small scales. Biological routes offer an advantage of inexpensive operating conditions at ambient temperature and pressure. Yet despite the long known existence of potential biocatalysts, biological conversion of gas to fuel (GTL) is still at early stages of research and is yet to show promise of commercialization. Two major hurdles in feasibility of attempted biological routes of gas fermentation are use of pure culture and toxicity of the end product [5–7]. The use of pure cultures needs initial sterilization to eliminate contamination at the time of installation and maintenance of sterile conditions throughout the operation to maintain constant activity; this significantly increases the process cost. The use of pure cultures, therefore, is only suitable for high-cost end products.

While scientists struggle with issues of mass transfer, high product toxicity, and low specific productivity [6] to create a capability of gas fermentation, huge reserves of biomethane and methane sinks in nature [8] tell us that viable solution and raw material exist. Easier solution, therefore, is to try to incorporate natural phenomenon much in the way it operates in nature and attempt to make the process as passive as possible (Fig. 1).

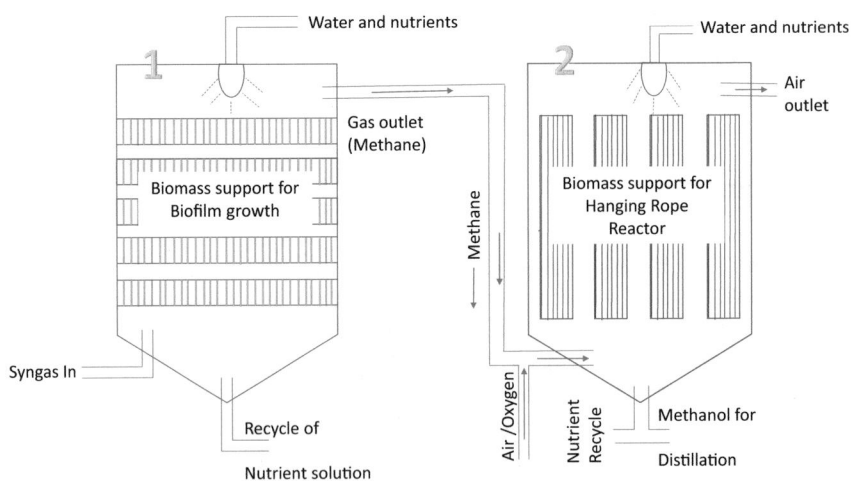

Fig. 1 Two-stage reactor: in the first stage: syngas to methane, and in the second stage: methane to methanol

2 Design: Two-Stage Reactor: Syngas → Methane and Methane → Methanol

2.1 Rational for Two-Stage Reactor

Micro-organism capable of converting CO_2 –CH_3OH is not yet known; however, distinct microbes which reduce carbon dioxide to methane and those which oxidize methane are well documented and have been proven to play an important role in maintaining C cycle. With an objective of mimicking this natural phenomenon and requirement of contrasting growth environment for the employed microbes, a two-stage reactor for carbon dioxide reduction in the first reactor (anaerobic) and oxidation of produced methane (aerobic) in the second reactor was designed.

2.1.1 Biocatalysts

- Hydrogenotrophic methanogens are anaerobic archae which consume CO_2 as its C source and hydrogen as the electron donor and reduce CO_2 to CH_4. These methanogens play a significant role in stabilization of the biogas plant and are known to be present in high concentrations in digested leaf biomass [3, 9]. Digested biomass hence becomes an easy source of these microbes for the methane-producing reactor.
- For the second reactor (methane-oxidizing reactor), aerobic methanotrophs would be employed. However, CO_2 is the natural metabolic end product of these microbes, while product of our interest is CH_3OH, first intermediary product of metabolic pathway. This necessitated the employment of chemical inducers to block the metabolic pathway after methanol production and supplements to compensate for the incurred energy loss from the arrested metabolism.

Consortia of these microbes would be enriched on the grown biofilms instead of attempting to obtain pure cultures to enable robust biofilm activity and lower the operating cost (maintaining pure culture is cost ineffective, and cultures are highly sensitive to contamination and growth conditions).

2.1.2 Biofilm-Based Reactor

High activity, shock tolerance, longer retention, self-regulating mechanism of biofilms is a common choice of survival in nature [10, 11]. Close association of microbes in biofilms or in granules enabled faster treatment of high strength waste water and is becoming preferred choice for waste water treatment reactors. Biofilms as the choice of microbial population distribution in the reactor also opens the possibility to minimize the requirement of liquid as the gas transferring medium to microbes, by enabling operation with a very thin film of liquid (just enough to ensure microbial sustenance). This should help in circumventing mass transfer limitations of poorly soluble gases. With these considerations, a biofilm-based reactor was envisioned.

Biomass Support of Choice

Biological biomass as biofilm support is known to produce high activity thin biofilms [9, 12]. Composition of biomass support also allows its use as food reserves in period of starvation, conferring enhanced shock tolerance.

- Spent biomass from herbaceous leaf based biogas plant can serve as biomass support and easy source of hydrogenotrophic methanogens. It is known to have high concentration of strongly adhered hydrogenotrophic methanogens (microbe of interest for methanation reactor) [10] which make major contribution (>90%) to produced methane [3]. Preparation time of methanogen enriched spent biomass is 15–30 days. [9]. This reduces lag (start-up period of anaerobic biofilms 3–9 months [13]), makes start-up inexpensive and biogas produced during the preparation phase (up to 0.54–1.09 l biogas/l biomass occupied volume/d after reaching steady state [9]) adds to energy economics of the entire process. Employment of biomass support at this stage begins to give >89% carbon dioxide conversion to methane in six days (4.49 ml CH_4/g biomass/d [CST, unpublished study, 2018]).
- Aerobic bacteria have high growth rate. The growth of these bacteria needs to be regulated to prevent problems of high slough off, blockage of gas diffusion paths affecting substrate access to bacteria and interfering in steady continuous operation of the reactor. Tough to degrade lignin rich coir fiber can be used as biological biomass support to regulate bacterial growth [12]. To further maintain the thickness of the biofilm, coir ropes would be packed vertically, which would have a provision of slow rotation of the ropes (optional), to enable easy sloughing off of the excess biofilm.

2.2 Operating Parameters

- Both the reactors would be operated at near ambient temperature and pressure. The active parts of the reactor would be gas flow regulators, water/nutrient sprinkling and re-cycling mechanism, rope/biofilm rotation system (in the case of methane fermentation reactor).
- Gaseous substrate would be fed at uptake rate of microbes to maintain a constant diffusion gradient and maximum conversion in single pass to ensure high product purity in effluent (gas in case of methanation reactor and liquid in case of methane fermentation reactor).
- Spent herbaceous leaf based biomass provides the nutrients needed to sustain the growth of anaerobic bacteria, eliminating the need to provide synthetic media.

Water sprinkling system of the reactor would be operated at rates which ensure presence of only thin layer of water (<1 mm) on the biofilm. The slower the flow rate of water, the less would be the bacterial washout. If the need to provide extra nutrients is deemed necessary, secondary treated waste water, a cheap alternate to expensive synthetic media would be used.

- Methane fermentation reactor needs to be provided with certain chemicals to stimulate the bacteria for methanol production. Such a synthetic media would contain chemical inducers (e.g., EDTA, NaCl etc.) to block the metabolic physiology of bacteria after methanol production, Cu to induce the cell to produce higher concentration of methane oxidizing enzyme (MMO) for high conversion rates of methane to methanol. Because of interrupted physiology, these cells are starved of reducing equivalents and hence would also need sodium formate like supplements of reducing equivalent [6, 14]. Regulation of nutrient laden water sprinkler in the case of second reactor needs extra consideration because of high toxicity of methanol [6]. Flow rates need to be slow enough to allow nutrient and substrate uptake (bacteria can uptake only the dissolved methane) but fast enough to prevent accumulation of toxic concentration of methanol.

3 Foreseeable Challenges

- In biofilm-based reactor fed with gaseous substrate, orientation of biofilm packing is of critical importance to prevent short circuiting and/or diffusion barriers which can severely affect reactor performance. Understanding of biofilm growth behavior a-priori is needed to engineer biofilm packing for maximum activity for periods matching half-life of the biomass support.
- Blocking the pathway at first stage to obtain methanol brings the energy kinetics of the cell under threat and greatly retards the growth potential [5, 6, 15]. Maintenance of cell at this critical stage, makes the controlling parameters stringent and the process expensive.
- Trade-off between building gradual tolerance for methanol from early stages of biofilm formation [inducing cells to produce methanol retards the growth (from 3 to 48 h)] at the cost of longer start-up period vs short start-up period at cost of incurring shock to mature biofilm.
- When bacteria are allowed to undergo complete metabolic pathway, the growth rate is as high as 3–6 h [15], this can allow fast biofilm build up. But it also means bacterial biofilm would experience significant shock when later it is operated under chemically controlled environment for methanol production. Drop in methanol production, possible extent of damage to the biofilm and required recovery time in such a case is not known, but is a serious factor for careful consideration. Long start-up period implies longer time of substrate requirement without desired end product concentration, an investment which would incur significant cost and increase the payback time.

4 Conclusion

The biological approach of waste management is environmentally benign and less expensive. It opens an opportunity of better waste management even at small scales, contributing in implementing the vision of decentralized waste management. Although there are serious challenges to be overcome for the success of this route, especially in the case of methane fermentation reactor (mentioned earlier in the paper), if successful, it can make waste management more efficient and lucrative for both developed and developing countries.

Biomethanation and methane fermentation reactor can be integrated with biogas plants and/or gasification plants. Here, methanation reactor can be used to increase methane concentration of the biogas either as a stand- alone reactor or integrated with methane fermentation reactor to obtain methanol. With each stage of the reactor, the end product is concentrated in terms of energy density. Such a system as a whole would create a capability to bring different approaches of waste treatment under an umbrella of a common drop-in fuel (methane/methanol) as the energy-rich end product.

References

1. Mospi.gov.in. (2018). *Area and population—Statistical year book India 2017 | ministry of statistics and program implementation | government of India*. [online] Available at: http://www.mospi.gov.in/statistical-year-book-india/2017/171. Accessed October 5, 2018.
2. Joshi, R., & Ahmed, S. (2016). Status and challenges of municipal solid waste management in India: A review. *Cognet Environmental Science, 2,* 1139434.
3. Chanakya, H. N., & Malayil, S. (2012). Anaerobic digestion for bioenergy from agro-residues and other solid wastes—An overview of science, technology and sustainability. *Journal of the Indian Institute of Science, 92*(1), 111–144.
4. Sheets, J. P., et al. (2016). Biological conversion of bio-gas to methanol using methanotrophs isolated from solid-state anaerobic digestate. *Bioresource Technology, 201,* 50–57.
5. Duan, C. (2011). High-rate conversion of methane to methanol by Methylosinus trichosporium OB3b. *Bioresource Technology, 102*(15), 7349–7353.
6. Hur, D. H., et al. (2016). Highly efficient bioconversion of methane to methanol using a novel Type Methylomonas sp. DH-1 newly isolated from brewery waste sludge. *Chemical Technology and Biotechnology, 92*(2), 311–318.
7. Kim H. G. et al., 2010. Optimization of lab scale methanol production by Methylosinus trichosporium OB3b0 *Biotechnology and Bioprocess Engineering, 15*(3), 476–480.
8. Hanson, R. S., & Hanson, T. E. (1996). Methanotrophic bacteria. *Microbiological Reviews, 60*(2), 439–471.
9. Chanakya, H. N., Venkatsubramaniyam, R., & Modak, J. (1997). Fermentation and methanogenic characteristics of leafy biomass feedstocks in a solid phase biogas fermentor. *Bioresource Technology, 62*(3), 71–78.
10. Chanakya, H. N., Srivastav, G. P., & Abraham, A. A., 1998. High rate biomethanation using spent biomass as bacterial support. *Current Science* 1054–1059.
11. Costerton, J. W., Geesey, G. G., & Cheng, K. J. (1978). How bacteria stick. *Scientific American, 238*(1), 86–95.

12. Chanakya, H. N., & Khuntia, H. K. (2014). Treatment of gray water using anaerobic biofilms created on synthetic and natural fibers. *Process Safety and Environmental Protection, 92*(2), 186–192.
13. Lauwers, A. M., Heinen, W., Gorris, L. G., & Van Der Drift, C. (1990). Early stages in biofilm development in methanogenic fluidized-bed reactors. *Applied Microbiology and Biotechnology, 33*(3), 352–358.
14. Hwang, I. Y., et al. (2014). Biocatalytic conversion of methane to methanol as a key step for development of methane-based bio-refineries. *Journal of Microbial Biotechnology, 24*(12), 1597–1605.
15. Whittenbury, R., Phillips, K. C., & Wilkinson, J. F. (1970). Enrichment, isolation, and some properties of methane-utilizing bacteria. *Microbiology, 61,* 205–218.
16. Kalyuzhnaya, M. G., et al. (2015). Metabolic engineering in methanotrophic bacteria. *Metabolic Engineering, 29,* 142–152.
17. Le Mer, J., & Roger, P. (2001). Production, oxidation, emission and consumption of methane by soils: a review. *European journal of soil biology, 37*(1), 25–50.

PCDD/PCDFs: A Burden from Hospital Waste Disposal Plant; Plasma Arc Gasification Is the Ultimate Solution for Its Mitigation

Saikat Das, Abhijit Hazra and Priyabrata Banerjee

Abstract Standing on the twenty-first century, the world is suffering from critical problems; growth in population invokes simultaneous generation and accumulation of waste materials. The unscientific way of waste disposal causes severe diseases, responsible for environmental and ecological imbalance. One of the major issues is improper disposal of generated hospital solid waste (HSW); with growth in population, there is simultaneous growth in patients and as an artifact concomitant piled up of generated HSW. The general and widespread methods for waste management comprises land filling and incineration of hazardous HSW that causes harm to our habitat by GHGs emission and accelerates global warming. The presence of chlorinated compounds like polychlorinated dibenzo-dioxins and polychlorinated dibenzofurans (PCDD/PCDFs) as solid residue in incinerated fly ash reduces the fertility of soils and deteriorates their qualities with progression. The presence of PCDD and PCDFs for several years produces several life-threatened diseases like cancer. Emphasis is given to develop a technology that can be highly efficient for complete and safe disposal of hospital wastes, with significantly less emission of chlorinated compounds or gases in environment. Plasma arc-driven technology can be a unique solution for mitigating HSW disposal like burning problems. Disposal of HSW inside plasma generating chamber in the absence of air or oxygen at the temperature zone 3000–6000 °C significantly reduces the chances of formation of PCDD and PCDFs. In addition, secondary treatment in adjacent chamber of the integrated waste management plant is of prior importance. In summary, evolution of flue gas signifies the complete burning of pathogens at 1000–1500 °C. The rarefied combination of several system components, e.g., plasma hearth, secondary chamber, ammonia and

S. Das · A. Hazra · P. Banerjee (✉)
CSIR-Central Mechanical Engineering Research Institute (CMERI),
Mahatma Gandhi Avenue, Durgapur, West Bengal, India
e-mail: pr_banerjee@cmeri.res.in

S. Das
e-mail: mrsaikat.das.smit@gmail.com

A. Hazra · P. Banerjee
Academy of Scientific and Innovative Research at CSIR-Central Mechanical Engineering
Research Institute (CMERI), Mahatma Gandhi Avenue, Durgapur, West Bengal, India
e-mail: hazraabhijit94@gmail.com

© Springer Nature Singapore Pte Ltd. 2020
S. K. Ghosh (ed.), *Energy Recovery Processes from Wastes*,
https://doi.org/10.1007/978-981-32-9228-4_2

9

water scrubber, condenser, booster drive, etc., makes the integrated HSW disposal procedure safe and environment-friendly. The presence of chlorinated gases is significantly lower as per standards in India and signifies the practicality of plasma arc-based next-generation solution to HSW.

Keywords Hospital solid waste disposal · PCDD · PCDF · Plasma treatment for mitigation of hazardous materials · International society of waste management · Air and water

1 Introduction

Disposal and management of hospital waste following proper, scientific, and safe method without affecting environment is a big issue for every nation. Today's world is suffering from two major problems: one is the escalation of inhabitants, while the other, as a consequence of it, is the uncontrolled generation and improper disposal of wastes. With the rapid change at the environment, people are easily affected by different diseases which signify commendable lose in their immunity power. People become sick and the generation of hospital waste is being increased throughout the globe in the last few decades. More emphasis should be drawn in case of hazardous hospital waste generation, handling, and transportation as well as disposal methods toward maintaining a sustainable clean environment.

Different hazardous elements, pathogens, and chemicals make HSW not only poisonous to the human being but also to the environment [1]. Doctors, nurses, medical stuffs, and other medical personnel can be seriously affected by the contamination of hazardous hospital wastes; moreover, cattle, rag pickers, and ecology may get seriously harmed if they come in direct contact to such contagious hospital wastes. Considering hospital waste and household waste, some countries dispose such waste through landfilling messing with household municipal solid waste [1, 2]. Even some developed countries have strong regulating guidelines for hospital waste handling and disposal, though these are not properly disposed throughout [2]. The different elements present as adsorbable organic halogens (AOH), chlorine and different chlorinated components, chemicals used as disinfectant, and tension actives make the total waste more toxic in nature [1]. The presence of chlorinated components in generated hospital waste causes severe environmental degradation. Due to the improper handling of hospital waste, generated chlorinated components are stored in open land that pollutes surface water and groundwater and even causes air pollution by producing polychlorinated dibenzo-p-dioxins or polychlorinated dibenzo furans [3]. The presence of urine, chlorinated chemicals, and autoclave disinfectant chlorine causes higher accumulation of chlorine as hospital waste. Some of the well-known waste disposal technologies as incineration produce harmful polychlorinated dibenzo-p-dioxins or polychlorinated dibenzofurans as an outcome of the disposal [3]. These PCDD or PCDFs are highly toxic in nature because they are highly stable in the atmosphere [4]. The effect of pollutants like PCDD and PCDF can be very harmful toward the environment. Such chemicals were used by the US army during Vietnam War, and as a consequence, the cultivated land, general people, and cattle were affected tremendously and throat cancer, lungs cancer, etc., were also endemic in nature.

2 Incineration—A Well-Known Method for Hospital Waste Disposal

This is a well-established technology in the entire world for the disposal of HSW. The waste materials are first put inside a closed chamber where it is burned or combusted in the presence of flammable oxygen. The feedstock is mainly hospital solid waste which contains pathogens, human blood, body fluids, amputated body parts, chemicals, disinfectants, medicines, stripped medicines, etc. Some hospitals use chlorinated compounds to disinfect hazardous, infectious hospital wastes in auto-clave process, and then, disinfected hospital waste is sent to the incineration chamber for further processing and final clearance. The main concern associated with the disposal of HSW through incineration is the incomplete combustion of waste materials and the gradual production of CO_2 and CH_4 as GHGs and also generation of polychlorinated dibenzo-p-dioxins (PCDD) and polychlorinated dibenzofurans (PCDFs) as flue gases. Generation of the GHGs causes severe problems like global warming, while PCDD and PCDFs remain unaltered in the atmosphere for a very long time and cause serious environmental degradation [5].

Mininnia et al. have developed a disposal system for hospital waste through incineration that comprises of rotary kiln furnace, followed by after burning chamber, air exhaust gas heat exchanger, water exhaust gas heat exchanger, dry reactor, bag filter, scrubber, and a chimney. The experiment was carried on varying eight different parameters, and temperature variations at three points as after burning chamber, water exhaust gas heat exchanger, and chimney are monitored and analyzed. After the kiln furnace, the temperature varies within the limit of between 910 and 1200 °C; in after burning chamber, the temperature further decreases in the range of 190–351 °C; in water exhaust gas heat exchanger, finally, the produced gas mixture is emitted through chimney where the temperature variation is about 70–96 °C (vide Fig. 1) [6].

The existence multiple chlorinated components in HSW enhances the release of PCDD and PCDFs as a final outlet. The authors observed in eight different observations the existence of PCDD/PCDFs is high in the after burning chamber (vide Fig. 2) which is in the range of 1670–8618 ng N/m^3, while the emission of such toxic element is about 77–894 ng N/m^3. The fly ash of incinerated hospital waste contains PCDD/PCDFs in the limit of 69–4915 $\mu g/m^3$. Thus, the existence of several toxic components in final outlet as fly ash and flue gas has become a major threat toward a clean eco-friendly HSW disposal [6].

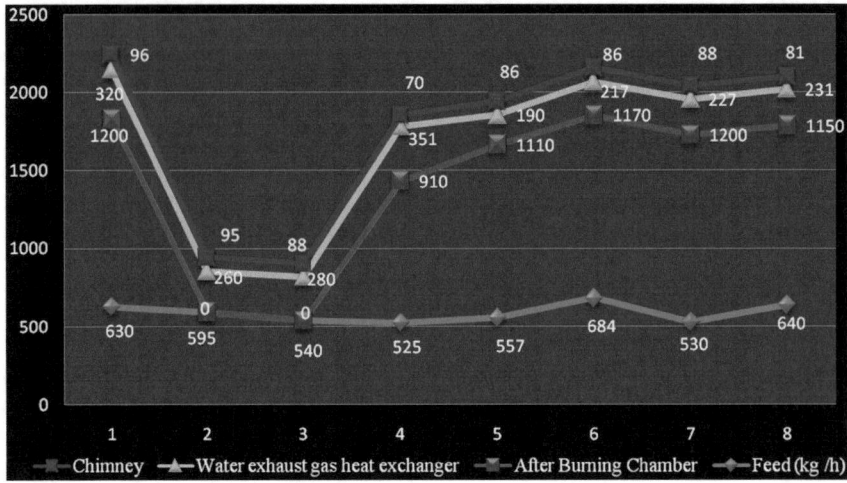

Fig. 1 Temperature variation in different components of incineration process [6]

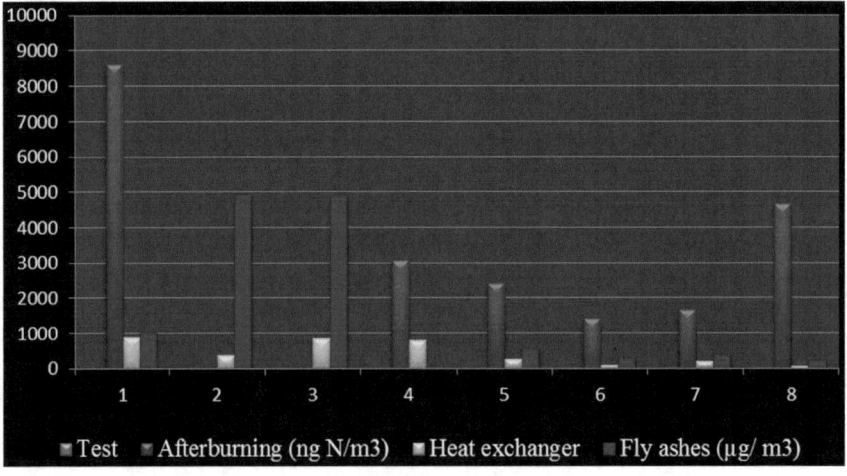

Fig. 2 Generation of PCDD/PCDFs inside different segments of incineration process [6]

Not only PCDD/PCDFs as toxic compounds, different polycyclic aromatic hydro-carbons are also produced by treatment through incineration of waste (vide Fig. 3). The carcinogenic polycyclic aromatic hydrocarbons are very much harmful for the environment. Through incineration of waste materials, the total generation of poly-cyclic aromatic hydrocarbons varies from 3377 ng N/m^3 to 11390 ng N/m^3 inside the after burning chamber of which carcinogenic compounds lie in the wide range of 25–1022 ng N/m^3. The presence of toxic effluents like polycyclic aromatic hydro-carbons gradually decreases in the next two parts: heat exchanger and stack. In

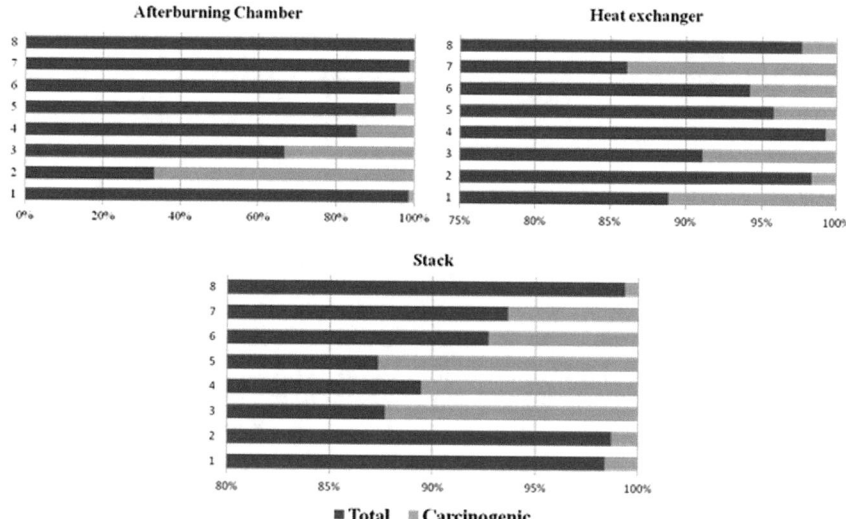

Fig. 3 Generation of polycyclic aromatic hydrocarbons inside different segments of incineration process [6]

the heat exchanger, the total polycyclic aromatic hydrocarbons vary in the range of 174–1669 ng N/m³. Carcinogenic compound that present in the final outlet is found to lie between the range of 4–66 ng N/m³. Total polycyclic aromatic hydrocarbons lie in the range of 763–4646 ng N/m³.

He et al. experimentally found that the emission of PCDD/PCDFs is huge in the final outlet of gas coming out of hospital waste treatment process through incineration. They recognized that contribution of furan as a toxic element is about 82%, while the contribution of dioxin is about 18% (vide Tables 1 and 2) where the main components are 1,2,3,4,6,7,8-HpCDF, 1,2,3,4,6,7,8-OCDF, 1,2,3,4,6,7,8-OCDD contributing about 7.35, 8.29, 2.60 ng I-TEQ N/m³, respectively. They also found that the dioxin was present in the range of 0.74 ng I-TEQ N/m³. The five main components together contribute about 80% of total PCDF emission. Among these five

Table 1 Generation of dioxin and furan and their various components through incineration process [7]

	Different elements of dioxin	% of presence in flue gas
Dioxin (18%) 0.74 ng I-TEQ N/m³ and Furan 82%	2,3,4,7,8-PeCDF	28.2
	1,2,3,4,7,8-HxCDF	10.5
	1,2,3,6,7,8-HxCDF	13.2
	2,3,4,6,7,8-HxCDF	16.6
	1,2,3,4,6,7,8-HpCDF	9.9

Table 2 Generation of furan and its various components through incineration process [7]

		Ng I-TEQ N/m^3	% of presence in flue gas
Furan	1,2,3,4,6,7,8-HpCDF	7.35	82%
	1,2,3,4,6,7,8-OCDF	8.29	
	1,2,3,4,6,7,8-OCDD	2.60	

2,3,4,7,8-PeCDF contributes about 28.2%, 1,2,3,4,7,8-HxCDF 10.5%, 1,2,3,6,7,8-HxCDF 13.2%, 2,3,4,6,7,8-HxCDF 16.6%, 1,2,3,4,6,7,8-HpCDF 9.9% [7].

Wang et al. experimentally found that the emission of PCDD/PCDFs is extensively increased at a very higher level if significant amount of PVC is found as waste material. The seventeen most toxic elements of PCDD and PCDFs are analyzed and found in four different observations that the percentage fraction of OCDD; the most toxic elements among these are found to lie in the range of 20–65%. Among the mixture of flue gas, 1,2,3,4,6,7,8-HpCDD and OCDD hold the highest percentage of the presence. It has been observed in four set of experiments in different working conditions [3].

3 Landfilling

Once landfilling was a reliable method for disposal of waste, where there is hardly any need for establishing any setup for disposal of waste. Waste materials are collected from several sources, and they are dumped on the landfilling sights for their natural degradation. As a consequence, the environment as well as cattle is badly affected by this, especially when there is huge pressure on land due to the growth in population. Hazardous waste as hospital waste when accumulated with MSW and disposed through landfilling causes a more generous hazardous pollution to the environment. The chlorine components directly come in contact to the ambience and can cause groundwater contamination and form several compound of dioxins and furans. Thus, landfilling became a severe threat to the complete disposal of HSW [8].

4 Plasma Arc-Driven Hospital Waste Disposal System for Treatment of Hospital Waste

There are basically two types of problems to the researchers when they deal with the different technologies for disposal of HSW. The first one is the huge accumulation of hospital waste: to find out a safe technology that can be fruitful to destroy completely the waste materials, and the other one is to maintain a proper ecological balance by emitting non-toxic effluents after disposal of HSW. The existing technologies such as incineration, autoclave, biological disinfection, and landfilling are either incapable of complete disposal of HSW or cause harmful toxic emission to the environment. At this situation, plasma arc-driven technology could be a viable solution to both the major problems.

Plasma is the fourth state of matter where ions, electrons, and neutral particles coexist at a higher temperature. The temperature in plasma generation zone is about 3000–6000 °C. At this very high temperature, the waste materials are completely disposed, the molecular bonds of the particles are broken, and they remain in the ionized form within the zone of plasma generation. After that, they constitute the lower bond gaseous molecules and get emitted. Organic bonds are incapable of remaining the same at the higher temperature.

5 Biomedical Waste Management Rules, 2016

The new biomedical waste management rules came on March 2016 by the Ministry of Environment, Forest and Climate Change, Government Of India which reveals that the safe limit of suspended particulate matter is 50 mg/Nm^3. The retention time for incineration of biomedical waste is 2 s. These rules are implemented to trim down the emission of dioxin and furan levels from the incineration chamber. Hypochloride-treated chemicals and biomedical waste are treated in combustion chamber causing toxic gas emission to the atmosphere owing to the presence of chlorine. It has been mentioned clearly in biomedical waste management rules that medical waste after being treated should not be mixed with MSW anyhow [9].

6 Formation of PCDD and PCDF by de Novo Synthesis Through Incineration of Biomedical Waste

PCDD and PCDF are generally formed due to the incineration of hospital waste; it is formed on the basis of some factors as operational temperature, the presence of chlorinated components in waste material, availability of oxygen, feeding procedures, retention time of HSW within combustion chamber, and amount of feedstock. Dioxins and furans are generally formed within the incineration chamber in the temperature range between 230 and 280 °C. PCDD and PCDFs are reformed at the temperature range of 150–200 °C [7]. These components are able to hold their stability in nature and these can be reformed in various temperature ranges. Through de novo synthesis, PCDD and PCDF are formed. The emission of toxic gases as PCDD and PCDFs in the atmosphere may cause malfunctioning of endocrine of human being that reduces the natural immune of body [10].

The generation of dioxin and furan is executed by three processes as high temperature synthesis, catalytic synthesis, and de novo synthesis. The high temperature synthesis occurs at the temperature range of 500–700 °C through the existence of chlorinated precursors. Catalytic synthesis occurs at the temperature range of 300–500 °C with or without the presence of chlorinated precursors. De novo synthesis occurs at the temperature range of 200–400 °C due to the presence of totally unburned carbon particles, chlorinated compounds, and some volatile components in ash [10].

7 Result and Discussion

7.1 Plasma Treatment for Hospital Waste Disposal

Plasma treatment for hospital waste disposal can be the real solution for treatment of hazardous hospital wastes completely and also it emits significantly less dioxin and furan into the atmosphere. The temperature generated within the plasma-generating chamber is in the range of 3000–6000 °C; at this high temperature, the formation of dioxin and furan is very much less because these toxic components are mainly generated in temperature range of 200–700 °C. Comparing with incineration process for treatment and disposal of waste, plasma arc-treated mechanism is more sustainable, causing less toxic emission to the environment.

7.2 Plasma Hearth Design for 1 Ton Per Day Hospital Waste Disposal

At CSIR-CMERI, a one ton per day hospital waste disposal plant has been designed. Plasma plume is formed within the primary plasma generation chamber in the absence of oxygen. Thus, very high temperature and dearth of oxygen restrict the formation of toxic elements like dioxin and furan inside this chamber. The salient feature of this technology is very high volume reduction (in the range of 95–99%). All organic waste components of HSW are gasified, and the inorganic parts such as glass and metals are liquidified and collected separately as slag. The produced gas is combusted in another chamber, while the slag material is used as an ingredient for construction of debris, bricks, pavement making materials, etc.

7.3 Combination of Plasma Electrodes and Plasma Torches for Plasma Arc Formation

Inside the plasma-generating chamber, plasma is produced by introduction of carrier gas argon. The setup has been developed and designed in such a way that a combination of electrodes and torches can be used simultaneously. Plasma arc is generated within the plasma-generating chamber where two opposite electrodes come in close vicinity with each other. Through the narrow passage of the plasma torch argon gas is sent which itself acts as plasma medium. There is a circular carbon block placed at the bottom of plasma-generating chamber which will act as an anode to the system. Cylindrical hollow carbon block acted as cathode generates plasma plume when these two opposite electrodes come closely and then drawn apart. When any external fluid is not used, the system acts as plasma electrode, while argon is used as plasma-generating gas, the system acts as plasma torch. Inside the plasma hearth, the temperature is so high that all organic compounds are being dissociated, and generated hospital waste is also ionized. The inorganic compounds are melted in high temperature plasma hearth, and there will be a definite percentage of vitrified non-leachable slag.

Priyabrata Banerjee et al./Sustainable Waste Management 2018

7.4 Secondary Treatment of Flue Gas Coming from the Outlet of Plasma Hearth Chamber

Hospital waste after being treated in primary plasma hearth, it generally gasifies 90–95% of the organic materials and the gas thus produced is treated within a secondary incinerator chamber. In the secondary chamber, the gas is incinerated in the presence of excess oxygen to make sure that there should not be any pathogen or harmful hazardous hospital waste remains in the final gas output. The temperature generated inside the secondary heating chamber is in the range of 1000–1500 °C. Through a spark plug ignition is created inside this chamber, air is supplied to the secondary treatment chamber through an air compressor.

7.5 Gas Analysis from the Outlet of Secondary Chamber

The gas output from the secondary treatment chamber is collected through a special setup and is finally analyzed in gas chromatography analyzer. Generally through the incineration procedure, generation of dioxin and furan is more than the plasma arc gasification process. Gas coming out from the outlet of secondary chamber is cleaned through a special system setup. The comparative analysis of WHO toxic emission standards with the toxic emissions from hospital waste disposal plant at CSIR-CMERI is analyzed (Tables 3 and 4).

Table 3 WHO report on accepted level of toxins in the environment [11], toxic element's presence in plasma arc-treated hospital waste disposal gas

Sl. No.	Structure	WHO-TEF
1.	2,3,7,8-TCDD	1
2.	1,2,3,7,8-PeCDD	1
3.	1,2,3,4,7,8-HxCDD	0.1
4.	1,2,3,6,7,8-HxCDD	0.1
5.	1,2,3,7,8,9-HxCDD	0.1
6.	1,2,3,4,6,7,8-HpCDD	0.01
7.	OCDD	0.0001
8.	2,3,7,8-TCDF	0.1
9.	1,2,3,7,8-PeCDF	0.05
10.	2,3,4,7,8-PeCDF	0.5
11.	1,2,3,4,7,8-HxCDF	0.1
12.	1,2,3,6,7,8-HxCDF	0.1
13.	1,2,3,7,8,9-HxCDF	0.1
14.	2,3,4,6,7,8-HxCDF	0.1
15.	1,2,3,4,6,7,8-HpCDF	0.01
16.	1,2,3,4,7,8,9-HpCDF	0.01
17.	OCDF	0.0001
18.	3,3′,4,4′-CB(77)	0.0001
19.	3,4,4′,5-CB(81)	0.0001
20.	3,3′,4,4′5-CB(126)	0.1
21.	3,3′,4,4′,5,5′-CB(169)	0.01
22.	2,3,3′,4,4′-CB(105)	0.0001
23.	2,3,4,4′,5-CB(114)	0.0005
24.	2,3′,4,4′,5-CB(118)	0.0001
25.	2,3,4,4′,5-CB(123)	0.0001
26.	2,3,3′,4,4′,5-CB(156)	0.0005
27.	2,3,3′,4,4′,5-CB(157)	0.0005
28.	2,3′,4,4′,5,5′-CB(167)	0.00001
29.	2,3,3′,4,4′,5,5′-CB(189)	0.0001

Priyabrata Banerjee et al./Sustainable Waste Management 2018

Table 4 Toxic element's presence in plasma arc-treated hospital waste disposal gas

Sl. No.	Parameter	Results (ng.TEQ)
1.	1 2 3 4 6 7 8-Heptachlorodibenzo-p-dioxin	<0.00024
2.	1 2 3 4 7 8-Hexachlorodibenzo-p-dioxin	<0.0024
3.	1 2 3 7 8 9-Hexachlorodibenzo-p-dioxin	<0.0024
4.	1 2 3 7 8-Pentachlorodibenzo-p-dioxin	<0.024
5.	2 3 7 8-Tetrachlorodibenzo-p-dioxin	<0.005
6.	1 2 3 6 7 8-Hexachlorodibenzo-p-dioxin	<0.0024
7.	Octachlorodibenzo-p-dioxin	<0.000015
8.	2 3 4 7 8-Pentachlorodibenzofuran	<0.0072
9.	1 2 3 4 6 7 8-Heptachlorodibenzofuran	<0.00024
10.	1 2 3 4 7 8 9-Heptachlorodibenzofuran	<0.00024
11.	1 2 3 4 7 8-Hexachlorodibenzofuran	<0.0024
12.	1 2 3 6 7 8-Hexachlorodibenzofuran	<0.0024
13.	1 2 3 7 8 9-Hexachlorodibenzofuran	<0.0024
14.	1 2 3 7 8-Pentachlorodibenzofuran	<0.00072
15.	2 3 4 6 7 8-Hexachlorodibenzofuran	<0.0024
16.	2 3 7 8-Tetrachlorodibenzofuran	<0.0005
17.	Octachlorodibenzofuran	<0.000015
	Total Dioxins and Furans	<0.01

7.6 Use of Water Scrubber, Ammonia Scrubber and Condenser for Gas Cleaning

Ammonia scrubber, water scrubber, and condenser are used as gas cleaning system. The gas coming out from the secondary treatment chamber is passed through these chambers. Toxic components as HCl vapor reacting with ammoniacal solution are getting separated from the emitted flue gas mixture. Thus, application of ammonia scrubber reduces the existence of toxic acid vapor in output gas mixture. The use of water scrubber or dry scrubber helps to reduce significantly ammonia vapor and other toxic gases from the gas mixture that has been generated after the treatment within secondary chamber. After that, the gas mixture is passed through a water-cooled condenser where the temperature of produced gas mixture is reduced, and then, finally it is released through a 30 m high chimney following the protocol of recent environment protection and pollution control rule, 2016.

8 Conclusion

Treatment of hazardous hospital waste is really a daunting challenge to the society due to its toxic and infectious nature. The treatment has to be channelized properly and must be scientific and specific that may not harm our environment. Plasma arc treatment in primary chamber and followed by secondary treatment inside secondary chamber would be the real solution of disposing of hospital waste. Emphasis has also been given to release non-toxic gases to the environment and to the atmosphere. Direct incineration of HSW emits toxic gases that are mainly PCDD and PCDFs. These chlorinated stable gaseous components are generated while hospital waste is treated in the temperature range of 400–1000 °C in incineration chamber. Dual treatment of HSW through plasma cracking chamber and secondary chamber emits significantly less amount of PCDD and PCDFs which remains within the emission standards of India's national toxic emission standards of HSW emission.

Acknowledgements Department of Science and Technology, Government of India-sponsored DST-TSG (vide letter no.-DST/TSG/WM/2015/459) (GAP-211712) project is hereby acknowledged. SD and AH are thankful to DST (GAP-211712) for their fellowship.

References

1. Carraro, E., Bertino, C., Lorenzi, E., Bonetta, S., & Gilli, G. (2016). Hospital effluents management: Chemical, physical, microbiological risks and legislation in different countries. *Journal of Environmental Management, 168,* 185–199.
2. Boillot, C., Bazin, C., Tissot-Guerraz, F., Droguet, J., Perraud, M., Cetre, J. C., et al. (2008). Daily physicochemical, microbiological and ecotoxicological fluctuations of a hospital effluent according to technical and care activities. *Science of the Total Environment, 403,* 113–129.
3. Wang, L. C., Lee, W. J., Lee, W. S., Chien, G. P. C., & Tsai, P. J. (2003). Effect of chlorine content in feeding wastes of incineration on the emission of polychlorinated dibenzo-p-dioxins/dibenzofurans. *The Science of the Total Environment, 302,* 185–198.
4. Zhu, F., Li, X., Lu, J. W., Hai, J., Zhang, J., & Xie, B. (2018). Emission characteristics of PCDD/Fs in stack gas from municipal solid waste incineration plants in Northern China. *Chemosphere, 200,* 23–29.
5. Hrabovsky, M., Hlina, M., Kopecky, V., & Maslani, A. (2017). Steam plasma treatment of organic substances for hydrogen and syngas production. *Plasma Chem Plasma Process, 37,* 739–762.
6. Mininnia, G., Sbrillib, A., Bragugliaa, C. M., Guerrieroc, E., Marania, D., & Rotatoric, M. (2007). Dioxins, furans and polycyclic aromatic hydrocarbons emissions from a hospital and cemetery waste incinerator. *Atmospheric Environment, 41,* 8527–8536.
7. He, J., Liu, T., Qiang, N., Li, Z., Cao, Y., Xie, L., & Zhao, Y. (2017). Emissions of PCDD/Fs in flue gas from a medical waste incinerator in Shanghai, *Earth and Environmental Science, 100.*
8. Havukainen, J., Zhan, M., Dong, J., Liikanen, M., Deviatkin, I., Li, X., et al. (2017). Environmental impact assessment of municipal solid waste management incorporating mechanical treatment of waste and incineration in Hangzhou, China. *Journal of Cleaner Production, 141,* 453–461.
9. Sheth, J. K. (2017). Salient features of bio-medical waste management rules, 2016. *Indian Journal of Community Health, 29.*

10. Wielgosinski, G., Namiecinska, O., Lechtanska, P., & Grochowalski, A. (2016). Effect Of selected additions on de novo synthesis of polychlorinated dioxins and furans. *Ecological Chemistry and Engineering S, 23*(2), 249–257.
11. Assessment of the health risk of dioxins: Re-evaluation of the Tolerable Daily Intake (TDI), WHO Consultation, May 25–29 1998, Geneva, Switzerland. http://www.who.int/ipcs/publications/en/exe-sum-final.pdf.

Catalytic and Non-catalytic Thermolysis of Waste Polystyrene for Recovery of Fuel Grade Products and Their Characterization

Rohit Kumar Singh, Biswajit Ruj, Anup Kumar Sadhukhan and Parthapratim Gupta

Abstract One of the important types of plastic extensively used in packing industry is polystyrene (PS). Having the lowest recycling rate among other categories of plastic waste, and not been included in the roadside recycling program, polystyrene is an attractive polymer waste for recovery of energy. PS mostly contains a large volume due to high air content in its products such as styrofoam. Being produced from petroleum product, it contains a large amount of energy which can be recovered by pyrolysis process which produces fuels like pyrolytic oil, combustible gases and char (carbon). In this study, zeolite (ZSM-5) catalyst was used during the catalytic pyrolysis process. Initially, TG analysis was performed to analyze the minimum degradation temperature of PS at four different heating rates and also to determine the effect of heating rate on its degradation. From the analysis, optimum temperature was selected based on time and temperature and was used as final temperature (500 °C) for batch pyrolysis. Ten percent of the catalyst is used with 200 gm of PS waste crushed in small size of 1–2 cm and then pyrolyzed. The main product from catalytic thermolysis was non-condensable gases 52%, pyrolytic oil 42% and remaining residue as char whereas non-catalytic process produces 78% pyrolytic oil, 14% gases and remaining 8% char. The products obtained from cracking in presence and absence of catalyst mainly constitutes aromatic, branched and cyclic hydrocarbons in the liquid phase, a high amount of hydrogen and C_1–C_4 hydrocarbons and some amount of CO and CO_2 due to the presence of oxygen in the raw material. The presence of catalyst during degradation increases the cracking forming a high amount of gaseous product whereas produces low range hydrocarbons (C_8–C_{16}) in oil phase as compared to non-catalytic pyrolytic oil which contains a high range of hydrocarbons (C_8–C_{25}).

Keywords Polystyrene waste · Thermolysis · Catalytic cracking · TGA

R. K. Singh (✉) · A. K. Sadhukhan · P. Gupta
Chemical Engineering Department, National Institute of Technology, M.G. Avenue,
Durgapur 713209, West Bengal, India
e-mail: ksinghmjpru@gmail.com

B. Ruj
Environmental Engineering Group, CSIR-Central Mechanical Engineering Research Institute,
M.G. Avenue, Durgapur 713209, West Bengal, India

© Springer Nature Singapore Pte Ltd. 2020
S. K. Ghosh (ed.), *Energy Recovery Processes from Wastes*,
https://doi.org/10.1007/978-981-32-9228-4_3

1 Introduction

High demand of plastics in various fields like agriculture, households, packing materials and various other applications, the plastics are produced in high quantities. An extensive research explains that plastics production is increasing by average of 5% every year on global basis since 1950 [1]. Therefore, the extensive usage of plastics also produced the waste which is further increasing day by day. In municipal solid waste, after food and paper, plastic is the third largest contributor [2]. About 90% of waste (MSW) in countries like India is deposited through land filling and plastics generally constitute about 7–10% of MSW [2, 3]. Polystyrene, one of the important types of plastic extensively used in packing industry, can be used to produce higher wt% of liquid oil to be used as an alternate fuel. The increase in demand of polystyrene clearly explains its uses and necessity of production. Many researchers have studied pyrolysis of polystyrene and co pyrolysis of polystyrene with other plastic types due to the capability of poly styrene to increase the yield of gaseous products [4]. Polystyrene is a particularly attractive synthetic polymer, of the major polymers, polysterene was excluded from recycling programs and has the lowest recycling rate among polymers [5]. There are several kinds of polystyrene produced in the industry for various purposes. These include EPS (Extended poly styrene), VPS (Virgin polystyrene), CPS (Polystyrene containers) and HIPS (High Impact Polystyrene). The difference lies in the composition and contaminants and additives added during the production of PS. For the reduction and management of plastic waste generally, primary recycling, landfilling and open burning are the main processes. Energy recovery methods are under extensive study where the plastics are converted to liquid oil through processes like cracking, carbonisation (pyrolysis) etc. Thermal cracking via pyrolysis is a suitable process for the tertiary recycling of PS [6]. Pyrolysis is a thermo-chemical process performed at high temperatures about 350–850 °C in absence of oxygen where the process breaks the bonds or carbon chain into smaller chains and sometimes monomers [2]. Pyrolysis of poly styrene can be done with and without catalysts, but there would be a difference in products and their composition. The major advantage of thermal process is the production of light hydrocarbons at low residence times and high temperatures. But during thermal cracking, the product obtained has limited commercial value and the product having a broad range of hydrocarbons is generally considered as the drawback of the process [6, 7]. Thermal pyrolysis was done by several researchers [8–10] for recovery of gasoline-range hydrocarbons. Onwudili et al. [10] performed a batch pyrolysis in pressurised autoclave reactor for 1 h at 300–500 °C and obtained a 97% liquid oil yield at a temperature of 425 °C which is considered optimized temperature and with a gas production of 2.5%. Similar study by Demirbas [3] showed that the liquid oil reduced to 89.5% when performed at a temperature of 581 °C during a batch process. To overcome the defects of thermal cracking and for a selective and better production of products catalytic cracking is used. Research scholars of Hamburg University [6] found that styrene concentrations in liquid oil can be reduced by using FCC catalyst. Experimental results showed that presence of FCC catalyst produced coke up to 20%

whereas non-catalytic pyrolysis produced only traces of coke. Besides, lower activation energy the influence of other catalysts on PS is less when compared to other plastics but HZSM-5 and Al_2O_3–SiO_2 because of higher acidity showed significant effect. It is observed that usage of HMCM-41 mainly produced ethyl benzene and benzene instead of styrene. Alkaline materials like magnesium and barium oxide are found to be selective towards styrene in PS cracking. There is a necessity for studying characteristics of individual plastics because of their importance in various fields. Since, the yield of high calorific plastic oil depends on type of process used this study provides a detailed investigation into catalytic and non-catalytic pyrolysis. In this study the effect of catalyst (HZSM-5) on PS pyrolysis was analysed and was compared with the thermal treatment process. The impact of catalyst on the product distribution, quality and quantity was discussed. In addition, the obtained gases and oil were also characterised for the enhanced effect of catalyst during the pyrolysis process.

2 Materials and Methods

2.1 Materials

The raw materials used for the catalytic and non-catalytic pyrolysis process was polystyrene waste in the form of EPS, CD covers, styrofoam, PS cutlery etc. All these waste were collected from the educational institute and their canteens located in Durgapur. The waste sample was first cut into a range of 1–2 cm pieces and then allow to rest in a hot air oven maintained at a temperature of 260 °C for 5 min so that the waste volume might reduce by melting. These melted-solidified waste is then used for the experimentation. Ultimate and proximate analysis was performed to determine the component present in it as shown in Table 1.

Table 1 Proximate and ultimate analyses of PS waste

Proximate analysis	Percentage (%)	Ultimate analysis	Percentage (%)
Moisture	0.28	Carbon	89.2
Ash	0.84	Hydrogen	8.78
Volatiles	94.33	Nitrogen	0.01
Fixed carbons	4.55	Sulphur	0.00

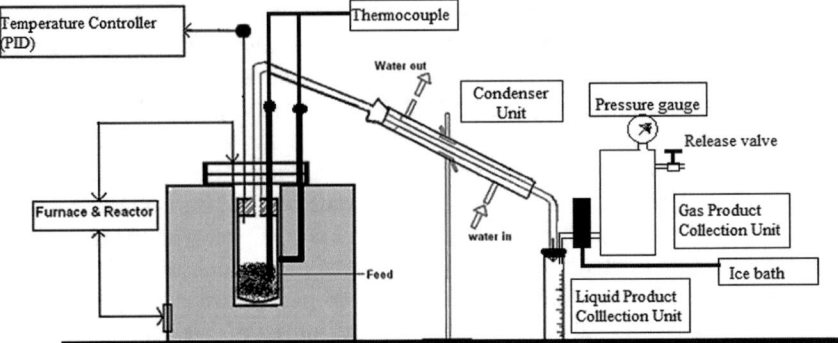

Fig. 1 Schematic of pyrolysis set-up

2.2 Experimental Set-up and Procedure

Pyrolysis experiments were carried out in an unstirred stainless steel reactor in a semi batch operation with a large amount to represent a real upscale process. The heating rate was at fixed at a rate of 20 °C min^{-1} during the operation. The experimental set-up is presented in Fig. 1. The set-up comprises of pyrolysis section condensation section, oil collection and gas collection units. The reactor is filled with waste and then purged with nitrogen for a period of 20–30 min at a flow rate of 200–300 ml/min for removing any O_2 from the reactor as well as from all other collection units attached with reactor. The reactor was heated up to a temperature of 500 °C in presence and absence of the catalyst (ZSM-10%). During degradation period, vapour leaving the reactor was passed through the condenser for gas-liquid separation; liquid pyrolytic oil was collected in the oil collection unit. Remaining non-condensable gases were passed through a water bath and collected in the gas collection tank.

2.3 Analytical Techniques

PS waste was analysed by CHNS analyser which compiles with the ASTM D5373 standard for elemental analysis of fuels. The proximate analysis for ash and moisture content are analysed according to D3174-82 and D3173-85 ASTM standards. A Thermo gravimetric analyser (make: SHIMADZU; model: DTG-60/60H) TG/DTA apparatus was used to analyse the thermal degradation behaviour of PS waste during pyrolysis reaction. The runs were performed at slow heating rate of 5, 10, 20, and 40 °C min^{-1}. Samples were flushed with Nitrogen at a flow rate of 0.2 L min^{-1} to ensure inert atmosphere. The experiments were repeated three times for each sample and the reproducibility was found to be within ±3%. Non-Condensable gases were analysed by Gas chromatograph (Model: GC-1000) from Thermo Fisher Scientific. The configuration for the Gas chromatography is shown in Table 2. The obtained oil

Table 2 GC configuration for PS waste pyrolysis sample analysis

Parameters	Analytical conditions	
Detector	TCD	FID
Column (Chemito make)	Spherocarb	Porapak-Q
Length	72 in. (6 ft)	96 in. (8 ft)
I.D.	1/8 in.	1/8 in.
Mesh size	80/100	80/100
Carrier gas	Argon	Argon
Flow rate (carrier gas)	30 ml/min	30 ml/min
Max. heating range (column)	225 °C	250 °C
Oven heating	50 °C	70 °C
Injector temperature	100 °C	225 °C
Detector temperature	150 °C	250 °C
Gas detection	H_2, N_2, CO, CO_2	CH_4, C_2H_4, C_2H_6, C_3H_8, C_4H_{10}
Sample injection	1.0 ml	0.2 ml

was also analysed using GC-MS for the determination of the major compounds as well as the distribution of hydrocarbons based on the carbon number distribution.

3 Result & Discussion

3.1 Thermogravimetric Analysis

TGA analysis of the PS waste is shown in Fig. 2a, b. The figure shows the different rate of degradation of PS waste. It was analysed that on increasing the heating rate, the degradation shifts towards high temperature and a rapid degradation was observed at high heating rates. In addition, it was observed that on increasing the heating rate the conversion increases with increase in amount of solid residue. Maximum conversion at all heating rate was obtained around 380–520 °C during which the maximum degradation takes place [7, 9]. The heating rate of 20 °C/min was used for the process of pyrolysis. During the process the products obtained weighted and yield was measured and gas was measured generally by difference.

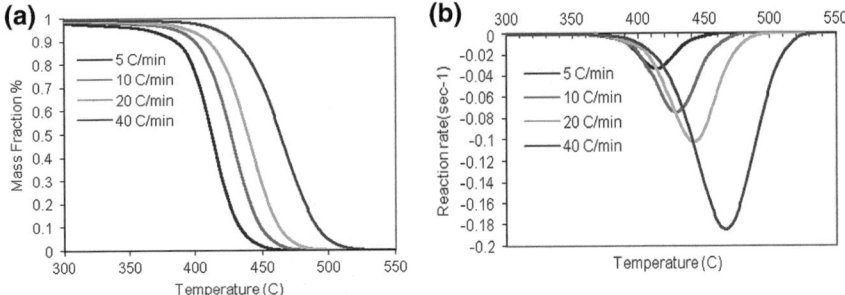

Fig. 2 **a** Mass fraction versus temperature. **b** Rate of reaction versus temperature

3.2 Effect of Catalyst on Product Yield Distribution

The product distribution for catalytic and non-catalytic pyrolysis of PS at a process temperature of 500 °C is shown in Fig. 3. It can be seen that there is a high impact of catalyst on the oil and gas yield whereas very little effect on the residue is observed. During the thermal pyrolysis process the degradation of long chain hydrocarbons is more prominent in medium length hydrocarbons causing higher formation of liquid phase product while there is also small production of short chain hydrocarbons which adds to the gas phase product [9, 10, and 12]. The thermal degradation of PS produces 78% pyrolytic oil having a density of 0.875 g/cm^3 whereas the gas phase product is 14% with the remaining as residual char. This residue is a good source of carbon as it contains a large amount of fixed carbon. In case of the catalytic pyrolysis the presence of catalyst increases the reaction rate as well as helps to break down the long chain hydrocarbons into short chain hydrocarbons and into monomer in a large quantity. The rapid cracking results into formation of lowered hydrocarbons as well as a large amount of free hydrogen atoms in the gas phase which results in higher gas phase product [9]. The catalytic process produces an enhanced gas phase product

Fig. 3 Product yield distribution for catalytic and non-catalytic pyrolysis of PS at 500 °C

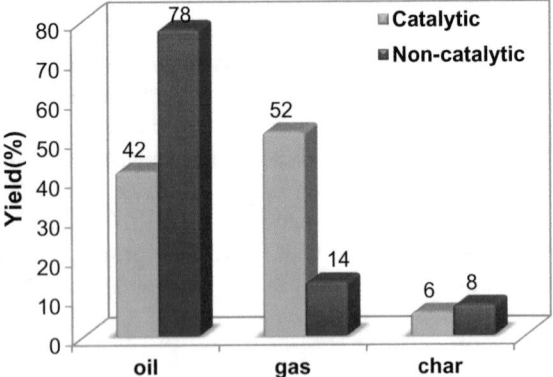

with 52% recovery whereas the oil phase product is decreased to a maximum of 42% oil yield. Although it has been observed that the oil obtained have a lower density and low range hydrocarbons fraction similar to gasoline which will be further discussed in Sect. 3.4.

3.3 Characterization of Pyrolytic Gas

The pyrolytic gases from both the catalytic and non-catalytic process which were obtained from the experiment performed at temperature of 500 °C for PS waste. Two detectors TCD and FID were used to determine the component presents in gases. IRIS 32 software was used to analyse the peaks provided by the chromatogram. Results from both detectors were compared and reported in Fig. 4. It is evident that the effect of catalyst on the PS degradation produces more low range hydrocarbons such as methane, ethane and butane in both quantity and quality. The concentration of methane and ethane has increased to 20.47 and 9.39% in catalytic pyrolysis from 13.75 to 4.01% which was obtained during the non-catalytic pyrolysis respectively whereas, the increase in concentration of propane was also increased from 0.74 to 3.70% during catalytic process. It was observed that the non-catalytic process favours the formation of hydrogen. CO_2, and higher hydrocarbons such as n–C_5 and n–C_6 which can also be observed from the Fig. 4. A higher amount of CO and CO_2 was also observed during the analysis which is the result of presence of air/oxygen obtained from the thermocol, styrofoam and EPS.

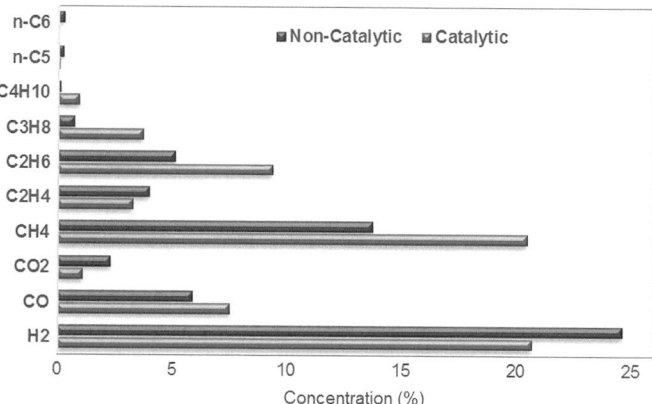

Fig. 4 Gas analysis for catalytic and non-catalytic pyrolysis of PS waste at 500 °C

3.4 Characterization of Pyrolytic Oil

The oil obtained was characterized based on the carbon number distribution using GC-MS. Table 3 represents the physical characteristics of the pyrolytic oils whereas Table 4 represents the fraction hydrocarbon fractions based on their carbon number. From Table 3 it can be easily observed that the catalytic oil have a lesser density as compared to the non-catalytic process whereas the viscosity was also found lower than the non-catalytic process. It can be observed from the analysis that the flash and fire point are nearby values whereas there is a high difference in the pour and cloud point of the pyrolytic oil. Also, the calorific value of the catalytic pyrolysis oil is comparatively higher due to the presence of low range hydrocarbons in the oil phase product [9–12].

It has been analysed that due to the presence of catalyst the cracking is rapid and more to form lower range hydrocarbons such as methane, ethane, propane etc. which are non-condensable whereas the low range hydrocarbon fraction concentration increases in the oil product (C_5–C_{13}). During non-catalytic pyrolysis the production of the oil phase product is higher in which the hydrocarbon ranging C_{14}–C_{24} have a higher fraction of 49.6% whereas in case of the catalytic process it reduces to 34.2% due to the higher cracking in presence of catalyst [12]. Catalyst allows the degradation of long chain hydrocarbons into shorter one which increases the C_5–C_{13} fraction in the non-catalytic process from 42 to 64% in catalytic process. Also it has been observed that non-catalytic produces a higher fraction of >C_{24} which is 8.4% whereas the same has been decreased to 1.8% during the use of catalyst. The most prominent compound was styrene [10–12] which is present in high quantity in both the cases whereas there are some other identified compounds presented in Table 5.

Table 3 Physical properties of catalytic and non-catalytic pyrolytic oil obtained from PS at 500 °C

Properties	Non-catalytic	Catalytic
Density (gm/cm^3)	0.875	0.781
Viscosity (cSt)	4.18	3.46
Flash point (°C)	36	35
Fire point (°C)	42	41
Pour point (°C)	−5	−14
Cloud point (°C)	−18	−27
Calorific value (MJ/kg)	39.871	43.11

Table 4 Hydrocarbon distribution during in pyrolytic oil from catalytic and non-catalytic pyrolysis process at 500 °C

Hydrocarbon range	Non-catalytic	Catalytic
C_5–C_{13}	42	64
C_{14}–C_{24}	49.6	34.2
>C_{24}	8.4	1.8

Table 5 Major compounds identified in oil fractions during catalytic and non-catalytic pyrolysis of PS at 500 °C

Major identified compounds	Non-catalytic (area%)	Catalytic (area%)
2-Methyl butane	3.8	5.4
2-Propenylidene-cyclobutene	3.19	4.48
Styrene	8.18	11.94
Cyclopentasiloxanedecamethyl	1.47	2.19
Benzene ethenyl- dimer	4.11	4.26
Tetradecanoic acid	2.57	2.98
Hexadecanoic acid	4.41	3.21
Benzene ethenyl- trimer	3.84	4.57
Squalene	2.14	1.19

4 Conclusion

This work represents the pyrolysis of PS in absence and in the presence of catalyst and the results were compared. From TG Analysis it is confirmed that increase in heating rate increases the degradation temperature whereas it also impacts on the residual fraction obtained. Presence of catalyst produces more gas phase products whereas absence of catalyst caused increased oil product during pyrolysis. There is no severe impact observed on the residual fraction (char). The physical analysis of oil shows that catalytic oil is superior in quality whereas quantity was highly affected by the presence of catalyst. The major component in the oil is styrene which is the monomer of the raw material. So it can be said that the process converts the waste into its raw material as well as produces products with high energy density. These products can be used for energy generation and hence energy can be recovered from the waste in sustainable way.

Acknowledgements The authors acknowledge the facility created by the FIST program of DST, Government of India for the sample analysis. Authors are also thankful to Director, CSIR-CMERI, Durgapur and Director-NIT, Durgapur for their support to carry out this research work.

References

1. Panda, A. K., Singh, R. K., & Mishra, D. K. (2010). Thermolysis of waste plastics to liquid fuel: A suitable method for plastic waste management and manufacture of value added products—A world prospective. *Renewable and Sustainable Energy Reviews, 14*(1), 233–248.
2. Singh, R. K., & Ruj, B. (2016). Time and temperature depended fuel gas generation from pyrolysis of real world municipal plastic waste. *Fuel, 174,* 164–171.
3. Demirbas, A. (2004). Pyrolysis of municipal plastic wastes for recovery of gasoline range hydrocarbons. *Journal of Analytical and Applied Pyrolysis, 72,* 97–102.

4. Williams, P. T., & Williams, E. A. (1999). Interaction of plastics in mixed-plastics pyrolysis. *Energy & Fuels, 13*(1), 188–196.
5. Mo, Y., Zhao, L., Chen, C. L., Tan, G. Y., Wang, J. Y. (2013). Comparative pyrolysis upcycling of polystyrene waste: Thermodynamics, kinetics, and product evolution profile. *Journal of Thermal Analysis and Calorimetry, 111*. https://doi.org/10.1007/s10973-012-2464.
6. Buekens, A. (2006). Introduction to feedstock recycling of plastics. In J. Scheirs, & W. Kaminsky (Orgs.), *Feedstock recycling and pyrolysis of waste plastics* (pp. 3–42). Hoboken: John Wiley & Sons.
7. Almeida, D., Marques, M. F. (2016). Thermal and catalytic pyrolysis of plastic waste. *Polímeros*, 26(1). http://dx.doi.org/10.1590/0104-1428.2100.
8. Butler, E., Devlin, G., McDonnell, K. Waste polyolefins to liquid fuels via pyrolysis: Review of commercial state-of-the-art and recent laboratory research.
9. Muhammad, C., Onwudili, J. A., & Williams, P. T. (2015). Thermal degradation of real-world waste plastics and simulated mixed plastics in a two-stage pyrolysis—Catalysis reactor for fuel production. *Energy & Fuels, 29*(4), 2601–2609.
10. Onwudili, J. A., Insura, N., & Williams, P. T. (2009). Composition of products from the pyrolysis of polyethylene and polystyrene in a closed batch reactor: Effects of temperature and residence time. *Journal of Analytical and Applied Pyrolysis, 86*, 293–303.
11. Dayana, S., Abnisa, F., Daud, W., & Aroua, M. (2016) A review on pyrolysis of plastic wastes. *Energy Conversion and Management, 115*. https://doi.org/10.1016/j.enconman.2016.02.037.
12. Liu, Y., Qian, J., & Wang, J. (1999). Pyrolysis of polystyrene waste in a fluidized-bed reactor to obtain styrene monomer and gasoline fraction. *Fuel Processing Technology, 63*, 45.

Energy Recovery from Tyre Waste Pyrolysis: Product Yield Analysis and Characterization

Rohit Kumar Singh, Biswajit Ruj, Anup Kumar Sadhukhan and Parthapratim Gupta

Abstract Yearly, out of 1 billion tyres, 40% are heavy-load vehicle tyre which was rejected or discarded to the dump sites where it piles up due to its non-degradable nature creating a problem for disposal. With the same rate of production, the problem is increasing day by day forcing the researchers to find alternatives for the disposal as well as energy recovery as these are a good source of carbon fuel. Methods such as open burning are not favoured because of high pollutant emission into the environment. Pyrolysis has emerged out as the best-suited option for the disposal as well as energy recovery. In this study, TG analysis was performed to determine the effect of temperature on the degradation of tyre waste in an inert atmosphere. Further, based on the analysis, heavy-vehicle tyres were pyrolysed in a temperature range of 550–800 °C for the recovery of valuable products such as pyrolytic as fuel oil, non-condensable gases as fuel gas and residue char as the carbon source. The pyrolysis process produces oil yield between 50 and 60% with a char yield of 32–38%, and the remaining were gases. The oil had a high calorific value of 34.743 kJ/Kg, whereas the evolved gas contains a heating value of 54.553 kJ/Kg. The gases obtained were analysed using gas chromatography showing a high percentage of hydrogen and C_1–C_4 hydrocarbon gases. The obtained char was analysed via SEM-EDX which confirms the presence of sulphur in high percentage. Also, the remaining char was activated using steam at 800 and 900 °C to increase the surface area for utilization as an activated carbon source and was determined by BET analysis.

Keywords Tyre waste · Pyrolysis · Fuel recovery · WTE

R. K. Singh (✉) · A. K. Sadhukhan · P. Gupta
Chemical Engineering Department, National Institute of Technology,
M.G. Avenue, Durgapur 713209, West Bengal, India
e-mail: ksinghmjpru@gmail.com

B. Ruj
Environmental Engineering Group, CSIR-Central Mechanical Engineering Research Institute,
M.G. Avenue, Durgapur 713209, West Bengal, India

© Springer Nature Singapore Pte Ltd. 2020
S. K. Ghosh (ed.), *Energy Recovery Processes from Wastes*,
https://doi.org/10.1007/978-981-32-9228-4_4

1 Introduction

Being a part of the urban stream, the generation of waste is increasing rapidly from different sources including the tyre waste which is also increasing day by day with an increase in population and a rapid boom in the transportation sector. After the utilization, the tyre has no end use and is generally considered as waste. In developed countries like Western Europe, USA, Japan, Australia, etc., the generation of tyre waste gave rise to stockpiling as these are non-biodegradable as well as considered as a loss of energy sources. These stockpiles add to the environmental pollution during open burning creating more toxic compounds causing health hazards. A similar scenario was also observed in developing countries like India [1–3]. Approximately, 1.5 billion tons of tyres are manufactured for their utilization in transportation vehicles such as a light vehicle, heavy vehicle, etc., and increasing [4]. About 0.6 million tons of scrap tyres is generated in India annually. Worldwide an estimated 13.5 metric tonnes of tyres were scrapped which includes all kinds of vehicle tyres such as cycle, motorcycle, cars/jeeps, tractors, trucks, and other heavy-load vehicles such as earth movers.

Different techniques were applied over time for the safe disposal of tyre waste as mis-handling of these results in a threat to human health and the environment. Waste tyres being non-biodegradable, landfilling is not a viable option for disposal, while incineration of these tyres results in the generation of a large number of dioxins and furans. Retreading, reclaiming are other options for the re-utilization of waste tyres [2, 3, 5]. An alternative solution for reducing the waste tyres with less environmental impact was pyrolysis which currently possesses huge attention from the researchers all over the world. Pyrolysis involves the thermal decomposition of the wastes at elevated temperature under an inert atmosphere like nitrogen. This process produces three forms of refined products such as condensable liquid (oil), non-condensable gases and the residue called as char in separated conditions. Condensable oil obtained from the pyrolysis of tyre waste is a good source of refinery feedstock for extracting high-value products, whereas it can also be used as furnace oil without any upgradation. The non-condensable gases obtained during the process have a higher calorific value and can be used for heating in the process itself. The residual char was obtained at the bottom of the reactor and had an economic value. These char or carbon black can be utilized for the activation process or as carbon black in remanufacturing of tyres. As all the products obtained have some economic value and can be used for energy recovery via appropriate technology [2, 3, 6], in this paper, we have carried out pyrolysis of heavy-vehicle tyre (Truck tyre) waste in a fixed bed reactor to obtain products like oil, gas and solid char at temperature ranges 550–800 °C at a heating rate 20 °C/min. The gaseous products were characterized using gas chromatography, whereas solid char were characterized using SEM-EDX.

2   Materials and Methods

2.1   Materials

In this study, heavy-vehicle tyre (truck tyre/lorry tyre) was used for the pyrolysis process acquired from the roadside scrap tyre vendors in Durgapur, West Bengal, India. The tread, shoulder and sidewall rubber were used for the analysis. The tyre was cut into small pieces of not larger than 2 cm and was used as a sample during pyrolysis. Ultimate and proximate analyses were performed to determine the component present in it as shown in Table 1.

2.2   Experimental Set-Up and Procedure

A fixed bed, non-stirred semi-batch, electrically heated reactor system was used for the process. The reactor is a cylindrically shaped reactor made from SS-304 with an internal diameter of 2 in. and of length 12 in. and can maintain a maximum pressure of 2 atm. The heating was done at a fixed rate of 20 °C/min during the operation. The experimental set-up is presented in Fig. 1. The set-up comprises of pyrolysis section condensation section, oil collection and gas collection units. The reactor is filled with waste and then purged with nitrogen for 10–20 min with a flow rate of 200–250 ml/min for removing any O_2 from the reactor as well as from all other collection units attached with the reactor. The reactor was heated within a range 550–800 °C. During degradation period, vapour leaving the reactor was passed through the condenser for gas–liquid separation; liquid pyrolytic oil was collected in the oil collection unit. Remaining non-condensable gases were collected in a gas collection tank after passing through a water bath and ice trap.

2.3   Steam Activation

During the activation of char obtained from the tyre waste pyrolysis, the char was again heated from room temperature to the required temperature of 800 and 900 °C

Table 1   Proximate and ultimate analyses of truck tyre waste

Proximate analysis	Percentage (%)	Ultimate analysis	Percentage (%)
Moisture content	1.61	Carbon	83.38
Volatile matter	63.70	Hydrogen	7.73
Ash content	5.43	Nitrogen	0.35
Fixed carbon	29.26	Sulphur	2.4

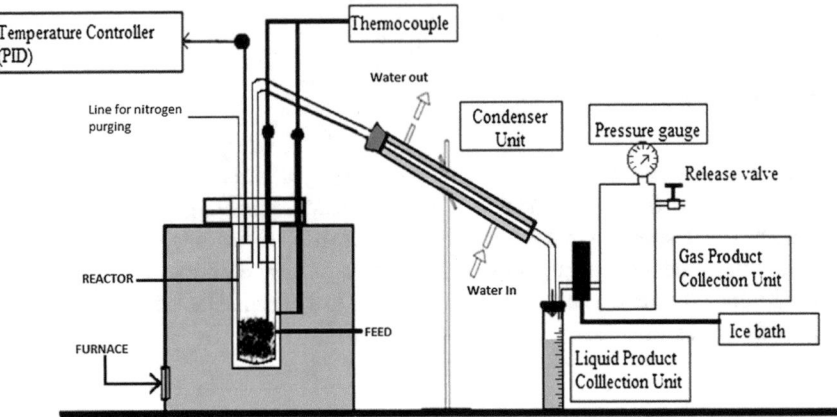

Fig. 1 Schematic of pyrolysis set-up

in an inert atmosphere obtained from continuous purging of nitrogen. After the temperature was reached, the sample was purged with steam and nitrogen was stopped inside the reactor for a period of 1 h. Further, the sample was cooled and tested for the surface area analysis.

2.4 Analytical Techniques

Truck tyre waste was analysed according to ASTM D5373 standard by CHN analyser for elemental analysis of the fuel. The proximate analysis for ash and moisture content is performed according to D3174-82 and D3173-85 ASTM standards. A thermogravimetric analyser (make: SHIMADZU; model: DTG-60/60H), TG/DTA apparatus, was used to analyse the thermal behaviour of tyre waste during pyrolysis reaction. The runs were performed at a slow heating rate of 5, 10, 20, and 30 °C min^{-1}. During the heating process, the environment is kept inert using nitrogen gas with a continuous flow at a rate of 40 ml/min. The experiments were repeated thrice to determine the reproducibility and found to be within ±2.3%. Non-condensable gases were analysed by gas chromatograph (Model: GC-1000) from Thermo Fisher Scientific. The configuration for the gas chromatography is shown in Table 2. Also, the sample obtained from the steam activation was analysed by BET analysis to determine the surface area of the char before and after activation.

Table 2 GC Configuration for tyre waste pyrolysis sample analysis

Parameters	Analytical conditions	
Detector	TCD	FID
Column (Chemito make)	Spherocarb	Porapak-Q
Length	72 in. (6 feet)	96 in. (8 feet)
I.D.	1/8 in.	1/8 in.
Mesh size	80/100	80/100
Carrier gas	Argon	Argon
Flow rate (carrier gas)	30 ml/min	30 ml/min
Max. heating range (column)	225 °C	250 °C
Oven heating	50 °C	70 °C
Injector temperature	100 °C	225 °C
Detector temperature	150 °C	250 °C
Gas detection	H_2, N_2, CO, CO_2	CH_4, C_2H_4, C_2H_6, C_3H_8, C_4H_{10}
Sample injection	1.0 ml	0.2 ml

3 Result and Discussion

3.1 Thermogravimetric Analysis

The degradation and reaction rate curves from TGA analysis of waste tyre truck are shown in Fig. 2a, b. The degradation of waste was analysed at four different heating rates of 5, 10, 20, and 40 °C/min to access the effect of heating rate. It was analysed that increased heating rate causes the degradation to shifts towards the high temperature and a rapid degradation was observed at high heating rates. In addition, on increasing the heating rate, the conversion also increases with an increase in solid residue. Maximum conversion at all heating rate was obtained around 350–450 °C during which the maximum degradation takes place. The heating rate of 20 °C/min was used for the process of pyrolysis. During the process, the products

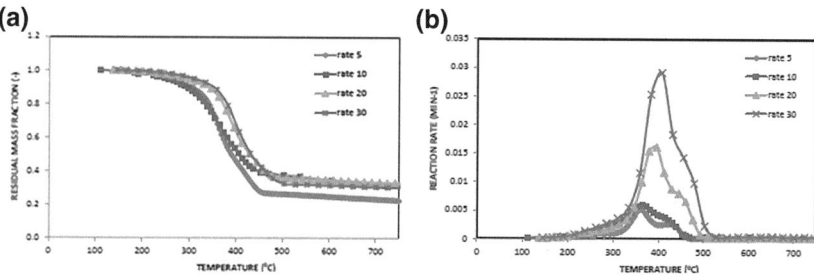

Fig. 2 **a** Mass loss fraction versus temperature, **b** rate of reaction versus temperature

obtained weighted, and yield was measured, while gas was measured generally by the difference.

3.2 Effect of Temperature on Product Yield

The product distribution obtained from the tyre for a temperature range of 550–800 °C for heavy-vehicle tyre waste at every 50 °C increment in temperature was measured and is shown in Fig. 3. At lower temperature, partial conversion results in a higher amount of solid contents whereas an increase in process temperature results higher conversion with low solid contents [6, 7]. Increased temperature results in higher condensable volatiles while further increase results in increased gas-phase product as depicted in Fig. 3. It was reported that the oil conversion increases up to a certain temperature, while further increase results in loss of condensable volatiles. The char content continually decreases with increase in temperature, whereas at higher temperature the gases yield increases [8, 9]. From the experimental investigation, 750 °C was considered as optimum temperature for the degradation of tyre waste with maximum oil recovery. Slow heating in the initial phase results in high residence time of volatiles in a reactor which results in higher gas production, whereas at higher temperature the secondary reactions play an essential role during cracking forming the high amount of C_1 and C_2 compounds [7–10]. At 750 °C, the amount of char increased as compared to 700 °C which can be explained by the further interaction of volatiles with char to form secondary chars.

Similarly, Williams et al. [7] pyrolysed waste tyre at varying heating rate and process temperature in a fixed bed reactor. They reported that high temperature and heating rate produces a more condensable product. At a heating rate of 5 °C/min, the condensable yield was 3% and 54.8% at a process temperature of 300 and 720 °C, respectively. Likewise, Mastral et al. [10] and Diez et al. [8] observed oil yields

Fig. 3 Product yield distribution versus temperature

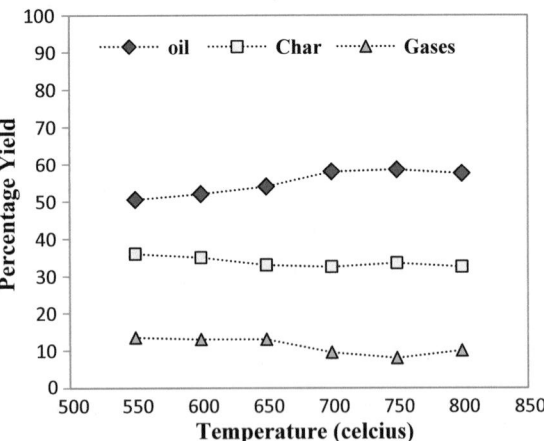

increased when tyres were pyrolysed at higher temperatures. In Rodriguez et al. [9] study, no significant difference in the energy value of the condensed oil was observed, whereas the yield increases with increase in process temperature from 300 to 700 °C from 4.8% to 38.5%. Laresgoiti et al. [11] obtained a significant increase in oil yield up to 500 °C with a maximum yield of 38.5% at 700 °C, while a similar yield of 38% was obtained at a processing temperature of 500 °C. The yield obtained at 400 °C was 24.8% [11].

3.3 GC Analysis of Pyrolytic Gas

During the pyrolysis process, the non-condensable gases evolved at an optimum temperature of 750 °C were analysed and reported in Table 3. During the degradation, the higher amount of SBR and NR results in the higher formation of C_1 and C_2 compounds as observed [6]. During the degradation process, olefins were evolved which goes under dehydrogenation reaction followed by Diels–Alder reaction resulting in a high amount of H_2 and methane. Further, the secondary reactions during pyrolysis also favour the formation of high gas-phase products [7, 10].

3.4 Characterization of Pyrolytic Char

3.4.1 Proximate and Ultimate Analyses of Tyre Char

The characteristic properties of the obtained char can be defined by the ultimate and proximate analyses and were provided within Table 4. The presence of volatiles in char further confirms the deposition of volatile compounds due to the secondary interaction of char with volatiles. The presence of synthetic rubber in the tyre waste results in higher ash contents, while higher carbon content was also observed [7–10]. Ultimate analysis reports a higher amount of carbon which is due to the presence of

Table 3 GC Analysis of pyrolytic gases by thermal conductivity detector (TCD) and flame ionization detector (FID)

S. No.	Compound name	Amount (%)
1	H_2	16.991
2	CO	4.880
3	CO_2	14.610
5	CH_4	9.092
6	C_2H_4	2.707
7	C_2H_6	12.791
8	C_3H_8	2.672
9	C_4H_{10}	4.011

Table 4 Proximate and ultimate analyses of tyre char obtained at 750 °C

Proximate analysis	Tyre waste char (%)	Ultimate analysis	Tyre waste char (%)
Moisture content	1.00	Carbon	86.13
Volatile matter	4.00	Hydrogen	5.23
Ash content	63.2	Nitrogen	6.45
Fixed carbon	31.8	Sulphur	2.04

unreacted carbon black used during the manufacturing of tyres, whereas the majority of the sulphur used for vulcanizing is retained in the solid residue. Also, a good amount of nitrogen was also reported. The amount of hydrogen and oxygen content in char has a significant impact on the combustion characteristics [2]. It was also analysed that a large amount of sulphur was present in the char and can be removed later. The presence of sulphur (85%) in char ensures less emission of sulphur to the environment.

3.4.2 SEM-EDX Analysis of Tyre Char

The surface structure was observed of the obtained char at different magnifications via SEM analysis and was shown in Fig. 4a–d. The analysis shows that the char surface has a high roughness which ensures high reactivity. The figures show that the high surface area was created during the degradation process while it was reported that slow heating results in high pores, whereas rapid heating results in the formation of cracks and channelling. For truck tyre pyrolytic char EDS or energy-dispersive X-ray spectroscopy was also performed to know the different components present in it with their concentration. Results are given below for pore diameter 500 μm. In this experiment, we use standard $CaCO_3$, FeS_2 and Zn for the detection of C, S, O and Zn, respectively. For different number of iterations, the results were obtained at different spots and results were reported in Table 5. The white spots in SEM images show the presence of sulphur. From the EXD analysis, it was observed that majority of the sulphur is retained in the solids (Fig. 5). It was reported that during the degradation process the sulphur reacts with the additives like ZnO and Fe_2O_3 and forms stable compounds as ZnS and FeS and retained in the solid state. Further, it was reported that with an increase in temperature the sulphur concentration decreases whereas the formation of H_2S in the gas phase increases [7, 9] (Table 5).

3.5 Tyre Char Activation

The char obtained at the optimum temperature (750 °C) was used for the activation process to increase the surface area and its utilization as activated carbon as it contains a high amount of fixed carbon. The char obtained directly has a surface area of

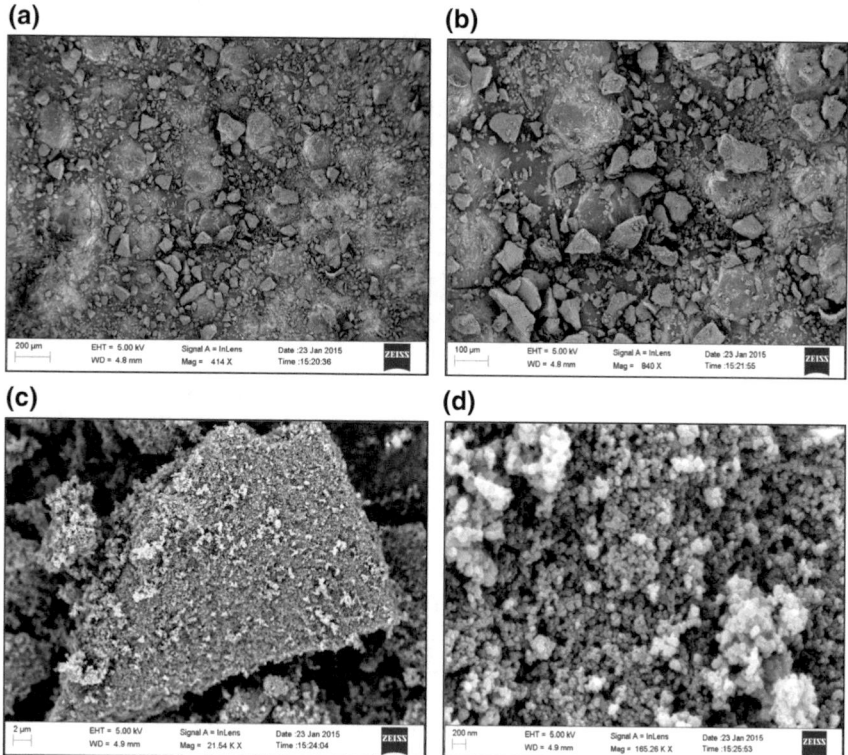

Fig. 4 SEM images of pyrolytic char obtained at 750 °C

Table 5 Component analysis of pyrolytic char through SEM-EDX analysis

Element	Point 1	Point 2	Point 3	Point 4	Point 5	Point 6
Carbon	96.30	73.29	91.57	100.00	95.29	91.87
Sulphur	1.21	6.55	2.62	00	1.15	1.97
Zinc	2.49	20.16	5.81	00	3.56	6.16

41.762 m^2/gm, whereas the steam treatment at 800 and 900 °C for 1 h increases the surface area to 225.399 and 408.745 m^2/gm, respectively. There is a reduction in weight % of 12–14% during the activation of char. At high temperature, steam activation had a high impact as at this temperature the steam penetrates more into the char particle to form more cracks which in turn increase the surface area and the pore volume.

Fig. 5 SEM-EDX image of pyrolytic char obtained at 750 °C at 500 μm resolution

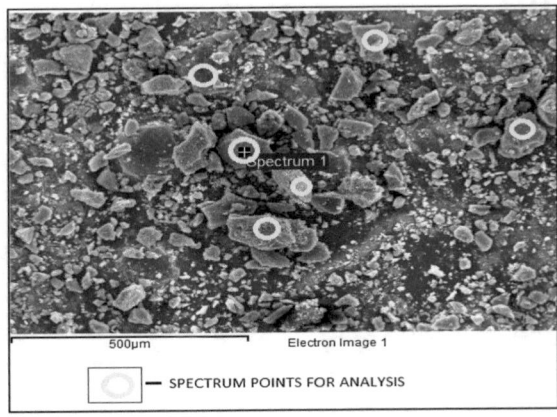

500μm Electron Image 1

☐ — SPECTRUM POINTS FOR ANALYSIS

4 Conclusion

During this study, tyre waste was successfully disposed of/reduced via pyrolysis process with the recovery of valuable products. The pyrolysis of the heavy tyre was done to obtain fuel oil and value-added products. The oil recovery % increases on increasing the temperature up to 58.5% at 750 °C which was more significant than the yield obtained by any other researchers. Also, it was observed that char percentage decreases on increasing temp, having most of the sulphur (85%) remains in the char. Also, a very little presence of sulphur in gas phase ensures the safety of the environment which has to vent off to the environment after burning.

Acknowledgements Authors are thankful to CSIR-Central Mechanical Engineering Research Institute, Durgapur, and NIT Durgapur for support to carry out this research work.

References

1. Recycling of tyres. www.tifac.org.in.
2. Swain, S. K. (2013). *Recycling of waste tyres: A possible option for deriving energy* (Thesis). http://ethesis.nitrkl.ac.in/5315/1/109ME0425.pdf.
3. Mahapatra, S. S. (2014). *Experimentation and evaluation of tyre pyrolysis oil* (Thesis). http://ethesis.nitrkl.ac.in/6363/1/110ME0300-10.pdf.
4. Bangladesh Bureau of Statistics. (2005). *Statistical year book of Bangladesh 2004* (24th ed.). Dhaka: Government of Peoples Republic of Bangladesh.
5. Jang, J. W., Yoo, T. S., Oh, J. H., & Iwasaki, I. (1998). Discarded tyre recycling practices in the United States, Japan and Korea. *Resource Conservation and Recycling, 22,* 1–14.
6. Rodriguez, I. M., Laresgoiti, M. F., Cabrero, M. A., Torres, A., Common, M. J., & Caballero, B. (2001). Pyrolysis of scrap tyres. *Fuel Processing Technology, 72,* 9–22.
7. Williams, P. T., Besler, S., & Taylor, D. T. (1990). The pyrolysis of scrap automotive tyres. The influence of temperature and heating rate on product composition. *Fuel, 69,* 1474–1482.

8. Diez, C., Martinez, O., Calvo, L. F., Cara, J., & Moran, A. (2004). Pyrolysis of tyres. Influence of the final temperature of the process on emissions and the calorific value of the products recovered. *Waste Management, 24,* 463–469.
9. Rodriguez, I. D. M., Laresgoiti, M. F., Cabrero, M. A., Torres, A., Chomon, M. J., & Caballero, B. (2001). Pyrolysis of scrap tyres. *Fuel Processing Technology, 72,* 9–22.
10. Mastral, A. M., Murillo, R., Callen, M. S., Garcia, T., & Snape, C. E. (2000). Influence of process variables on oils from tire pyrolysis and hydropyrolysis in a swept fixed bed reactor. *Energy & Fuels, 14,* 739–744.
11. Laresgoiti, M. F., Caballero, B. M., de Marco, I., Torres, A., Cabrero, M. A., & Chomon, M. J. (2004). Characterization of the liquid products obtained in tyre pyrolysis. *Journal of Analytical and Applied Pyrolysis, 71,* 917–934.

Solid-State Gas Fermenter to Convert Syngas to Methane

Aastha Paliwal and Hoysall N. Chanakya

Abstract As we source segregate decomposables from inerts, gasification becomes attractive for quick treatment and disposal of MSW, while also raising critical question of what needs to be done with excess or locally unusable syngas. With SDGs being drivers, we quickly need to find alternatives to efficiently and sustainably convert MSW in fuels, fertilizers, and locally value-added products. Conversion to methane or methanol is a simple option. Gas fermenting bacteria are known for more than a century, yet the potential to use them as biocatalyst for commercial GTF (gas to fuel) is being recently investigated. Methanotrophs, methanogens, and acetogens are the major groups in focus, where potential use in methanol production, increasing in situ methane content in biogas and syngas fermentation to produce liquid fuel, is active area of research but has met limited success so far. Fairly established science of methane production through CO_2 reduction via hydrogenotrophic methanogens, high turnover rates, and significance of these microbes in biogas stabilization suggests the potential of this route in GTF production. This paper describes a biofilm-based reactor as a potential reactor design for syngas conversion to methane.

Keywords GTF · Reactor design · Biofilm reactor · Hydrogenotrophic methanogens · Biofuel · Syngas · Methane

1 Introduction

The end of twentieth century witnessed a significant convergence of different fields of science for the hunt of better fuel alternatives. With increasing dependence of the 'global community' on energy (at the center of which lies fossil fuel-based energy) and its role in determining the growth of nations [12], search for 'economic, renewable and environmentally benign' alternate fuels/energy sources still remains the research focus cutting across various domains of science and technology. Roots

A. Paliwal · H. N. Chanakya (✉)
Centre for Sustainable Technologies, Indian Institute of Science, Bangalore, India
e-mail: chanakya@iisc.ac.in

A. Paliwal
e-mail: aasthap@iisc.ac.in

© Springer Nature Singapore Pte Ltd. 2020
S. K. Ghosh (ed.), *Energy Recovery Processes from Wastes*,
https://doi.org/10.1007/978-981-32-9228-4_5

45

of urgent need of alternate fuel can be traced back to 'fuel crises of 1970s' which brought proponents like hydrogen and solar, methane and methanol at front.

While chemical routes of production: synthetic (thermochemical) or bio-based 'chemurgy,' were obvious from the research trends of the twentieth century, concept of 'sustainable technologies' in 1987 [9] attracted a lot of microbiologists, who until then had limited their role to investigating underlying bacterial physiology and remained relatively distant from the field of energy production and waste management. This involvement of microbiologists changed the dynamics of wastewater treatment, making 'biofilm' (stagnant/moving)-based reactors popular choice and opening new (biological) routes of 'GTL'—gas-to-liquid fuel production using microbes, a process commonly referred as 'gas fermentation,' 'bio-GTL.'

Despite recent advances in the direction of gas fermentation, commercial use of hydrogenotrophic methanogens for carbon sequestration and hydrogen storage as methane has still eluded scientists. Large methane reserves, catalytic route of production from syngas, and low chemical production/procurement cost could be possible reasons for the lagging interest in tapping this potential. However, rising environmental concerns and realization of the significance of decentralized technologies in waste management, especially for a country like India where the resource potential is spread, make gas fermentation to methane an attractive route. With source segregation of MSW, gasification of undigestible fraction offers fast conversion to syngas, and integration with bioconversion of this gas to methane opens a new route of tapping energy from waste. Biological production of methane from syngas renders the process economic, rapid with a small environmental and physical footprint even at small scales (10–200 ton per day). This route of waste-to-energy (W2E) conversion will deliver methane as the end product, for which infrastructure is well in place. Integration of gasification with biological methane production using hydrogenotrophic methanogen increases the resource base, allowing MSW, surplus agro-residues, linear plastics, etc., to serve as input feed. The time frame of entire reaction is <24 h.

2   Limitations of Current Gas Fermenting Reactors

Current routes of gas fermentation to fuel (for methanol, ethanol, acetate, etc.) face severe feasibility limitations due to

1. The mass transfer limitation of less soluble gas (solubility of hydrogen < 2 mg/l [11])
2. High operating costs associated with pure cultures
3. Long startup period (>3 to 9 months) of biofilm based reactors [10]
4. High operating cost of CSTR
5. Toxicity of the end product (e.g., methanol, ethanol) [7]
6. Chemical, physiological, and/or genetic manipulation needed to block metabolic process after formation of product of interest (knowledge gap) [8]
7. Low specific productivity of product [7].

We propose a solid-state gas fermenter as a potential bioreactor design for conversion of syngas to methane using hydrogenotrophic methanogen as biocatalysts.

3   Solid-State Gas Fermenter

Technology for converting waste to syngas is in place [6]. The question that arises then is whether we can find any naturally occurring biological phenomenon to convert surplus syngas to drop in fuels like methane and methanol. This requires overcoming the foreseeable limitations of gas fermentation and reduction of operating and production costs. Figure 1 represents a sketch of the proposed solid-state gas fermenter for this purpose.

3.1   *Biocatalyst: Hydrogenotrophic Methanogens*

Inocula from mesophilic biomethanation plants quickly lower the redox potential of the reactor system (≤ -250 mV) by rapid consumption of oxygen (micro-aerophilic bacteria of the consortia) and all other oxidized species. Redox potential of this order arrests the growth of aerobic bacteria, naturally enriching the anaerobic bacteria. Administration of carbon dioxide as the only carbon source further moves the enrichment process in favor of hydrogenotrophic methanogens. This self-regulating nature of consortia enhanced by single carbon source, CO/CO_2, renders the need of stringent process control redundant. We seek to exploit this natural tendency of anaerobic consortia to create a biological process that can work at low input energy and operate successfully even at small scales.

Fig. 1 Solid-state gas fermenter for conversion of syngas to methane

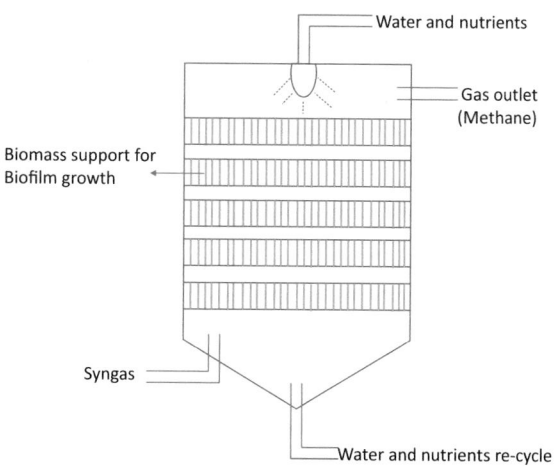

3.2 Biofilm-Based Approach

Ubiquitous presence of biofilms and associated removal of organic compounds (COD reduction) in the surrounding environment was described by Costerton [5]. He suggested the high treatment potential and high tolerance of biofilms to extreme environments (toxic material, high abrasion forces, competition, predators). Following his work, a large number of biofilm-based reactors (both stationary and moving) have been employed in wastewater treatment and the concept was adopted for gas fermentation reactors later where they conferred the advantage of retaining high microbial biomass with enhanced access to a gaseous substrate.

Natural (pumice stones), synthetic (PVC, styrene, PPE, etc.), and biological (coir fiber, ridge gourd fiber, rice husks, etc.) supports have been tried for a variety of treatment options [1]. Chanakya and Khuntia [1] demonstrated biological biomass support achieved similar COD reduction (>90%) of the wastewater despite being less dense (biofilm content on such supports was 4–10 times thinner than biofilms formed on synthetic support). The high activity of thinner biofilms on biological supports was hypothesized to result from low EPS, high bacterial cell concentration as opposed to high EPS, low bacterial cell concentration with synthetic support.

In category of biological biomass support, spent biomass from herbaceous leaf-based biomethanation plants (0.54–1.09 lbiogas/lbiomass occupied volume/d [4]) is an attractive alternative. In leaf digestion, >90% methane is contributed by hydrogenotrophic methanogens [2] which adhere to decomposing biomass in concentrations that can provide over 20 times higher gas conversion rates than occurring in the digester [4] and hence is an easy source of microbe of interest.

Digested leaf biomass residue after anaerobic digestion used as biomass support and source of methanogens provided rapid start-up, where 89.9% CO_2 conversion efficiency, producing 4.49 ml $CH_4/g_{biomass}/d$, was obtained in only 6 days [CST unpublished work, 2018]. Continuous disintegration of biomass support (half-life 93 days) might also contribute in methane production at some point. In an unpublished work at CST, 2018, sometimes ~125% conversion to methane was observed, and it is hypothesized that methane from continued disintegration was responsible for excess methane detected. This also suggests that potential of this biomass support to serve as food reserve rendering enhanced tolerance and fast recovery from starvation.

3.3 Operating Parameters

Solid-state methanogenic reactors have been tried for long to convert agro-residues and MSW to biogas without succumbing to problems of floating of feed. It is operated with very less liquid volume, enough to sustain microbes and appropriate nutrient supply, recycled to replenish the washed-out microbes (~40%) [3] and minimize nutrient requirement. Porosity and high surface area of decomposing biomass as well as a gradually descending film of digester liquid leaching down the biomass bed provide necessary conditions for effective diffusion of even poorly soluble hydrogen into hydrogenotrophic methanogens. This increases the accessibility of gas to methanogens, who have very high turnover rate for hydrogen, resulting in high reaction rates.

4 Conclusion

Integration of solid-state biofilm reactor with syngas generator is proposed to convert MSW to methane. Such plants would allow faster and high purity conversion of waste to methane and obviate need for various forms of rejects and inerts to be landfilled at high economic and environmental costs. With low water requirements, self-regulating thin yet highly active biofilms, methane can be generated from an otherwise low-quality syngas that has fewer local applications. Gas flow, water recycling, and sprinkling system being the only active part, the reactor is relatively passive. In future, we can look forward to adopt same principle for gas fermentation of methane to methanol using methanotrophs and integrate the two reactors to enable conversion of syngas to methanol in a two-stage reactor.

References

1. Chanakya, H. N., & Khuntia, H. K. (2014). Treatment of gray water using anaerobic biofilms created on synthetic and natural fibers. *Process Safety and Environmental Protection, 92*(2), 186–192.
2. Chanakya, H. N., & Malayil, S. (2012). Anaerobic digestion for bioenergy from agro-residues and other solid wastes—An overview of science, technology and sustainability. *Journal of the Indian Institute of Science, 92*(1), 111–144.
3. Chanakya, H. N., Srivastav, G. P., & Abraham, A. A. (1998). High rate biomethanation using spent biomass as bacterial support. *Current Science*, 1054–1059.
4. Chanakya, H. N., Venkatsubramaniyam, R., & Modak, J. (1997). Fermentation and methanogenic characteristics of leafy biomass feedstocks in a solid phase biogas fermentor. *Bioresource Technology, 62*(3), 71–78.
5. Costerton, J. W., Geesey, G. G., & Cheng, K. J. (1978). How bacteria stick. *Scientific American, 238*(1), 86–95.
6. Eliasson, A., Karstensson, J., Kronander, A., Karlsson, H. T., Hulteberg, C., & Svensson, H., (2015). Feasibility study of gasification of biomass for synthetic natural gas (SNG) production.
7. Hur, D. H., et al. (2016). Highly efficient bioconversion of methane to methanol using a novel Type Methylomonas sp. DH-1 newly isolated from brewery waste sludge. *Chemical Technology and Biotechnology, 92*(2), 311–318.
8. Kalyuzhnaya, M. G., et al. (2015). Metabolic engineering in methanotrophic bacteria. *Metabolic Engineering, 29*, 142–152.
9. Keeble, B. R. (1988). The Brundtland report: 'Our common future'. *Medicine and War, 4*(1), 17–25.
10. Lauwers, A. M., Heinen, W., Gorris, L. G., & Van Der Drift, C. (1990). Early stages in biofilm development in methanogenic fluidized-bed reactors. *Applied Microbiology and Biotechnology, 33*(3), 352–358.

11. Phillips, J. R., Huhnke, R. L., & Atiyeh, H. K. (2017). Syngas fermentation: A microbial conversion process of gaseous substrates to various products. *Fermentation, 3*(2), 28.
12. Yergin, D., & Bocca, R., 2012. Energy for economic growth. Report. Geneva, Switzerland.

Influence of Pyrogallol (PY) Antioxidant in the Fuel Stability of Alexandrian Laurel Biodiesel

P. Mohamed Nishath, K. Sekar, P. Mohamed Shameer, Kaisan Muhammad Usman, Abubakar Shitu, J. Senophiyah Mary and B. Dhinesh

Abstract The biodiesel quality can be maintained by the most significant criteria called 'Storage oxidation stability.' The chief technical blockage related with the commercialization of biodiesel is the poor oxidation stability. The investigation of present paper involves the effects of pyrogallol (PY) antioxidant which helps to maintain the thermal stability, accelerated stability, and storage stability for a long tenure by additive concentration. For different concentrations of PY, the C–H bonds and O–H bonds regions following the biodiesel oxidation variability can be characterized by the Fourier-transform infrared (FTIR) spectroscopy. The stability for oxidation can be increased by 95.67%, stability for storage by 15.42% and stability for thermal by 71.24% were obtained by adding up of PY at 950 ppm concentration (B100P3) enhanced with biodiesel which is in pure nature. Further concentration of antioxidant leads to the deterioration of hydrophilic and hydrophobic clusters formation which is characterized by the FTIR spectrum data. From the investigation, it is came to a conclusion that by dosing 950 ppm of PY antioxidant the Alexandrian laurel biodiesel could be accumulated over a long period of time.

Keywords Alexandrian laurel · Pyrogallol · Storage stability · Oxidation stability · FTIR · Thermal stability

P. Mohamed Nishath (✉) · J. Senophiyah Mary
Government College of Technology, Coimbatore 641013, India
e-mail: pmdnishath@gmail.com

K. Sekar
Deparment of Mechanical Engineering, National Institute of Technology,
Calicut, Kerala, India

P. Mohamed Shameer
Deparment of Mechanical Engineering, Faculty of Engineering,
V V College of Engineering, Tirunelveli, India

K. Muhammad Usman · A. Shitu
Deparment of Mechanical Engineering, Faculty of Engineering,
Ahmadu Bello University, Zaria, Nigeria

B. Dhinesh
Deparment of Mechanical Engineering, Mepco Schlenk Engineering College,
Virudhunagar, Tamil Nadu, India

© Springer Nature Singapore Pte Ltd. 2020
S. K. Ghosh (ed.), *Energy Recovery Processes from Wastes*,
https://doi.org/10.1007/978-981-32-9228-4_6

Nomenclature

FTIR   Fourier-transform infra red
PY   Pyrogallol
TGA   Thermogravimetric analyzer
T_{ON} Onset temperature

1   Introduction

Momentous attention of biodiesel is due to energy concerns like hike in fuel production cost, demand, or supply of fuels due to instability and harmful environmental impact due to fossil fuel burning [1]. In 2014, biodiesel production has been increased globally by 7.4% and among India biodiesel production is increased by 29.1% which is about 0.320 million tones than the previous year [2]. When compared to the year 2014 oil consumption level has been escalated by 1.6, 6.2, 2.6, 6.4, 1.8, 6.3, and 8.1% in Italy, USA, Brazil, China, Australia, and India by the year 2015 [3]. In USA, due to implementation of renewable energy, the NO_x emission has been declined by 50% [4]. The global energy crisis is mainly due to the emissions from the transportation sector [5]. The main degradation in biodiesel is due to contamination of foreign particle, thermal-oxidation decomposition, hydrolysis, and oxidation which will lead to change in fuel properties and hindering in engine performance [6, 7]. The term 'fuel stability' is mainly focused for the biodiesel degradation. The properties like thermal stability, oxidation stability, and storage stability were conversed for the stability of fuel [8]. The oxidation reaction can be stabilized by antioxidants which will react with peroxide radicals which lead to the liberation of antioxidant free radical, thereby inhibiting the process of oxidation [8, 9]. High value of heat liberation and high yield of oil can be obtained from Alexandrian laurel when compared to neem, jatropha, and karanja biodiesel sources [10]. Table 1 represents the properties of fuel blends.

From the study, it has been observed that the oxidation stability analysis in biodiesel shows 44.57% higher oxidation stability when 10% pentanol with adding up of 20% biodiesel [11]. Characteristics of diesel engine like performance, emission, and combustion can be improved by changing the operating parameters like compression ratio and injection timing [12, 13]. For characterizing the oxidation stability of various concentrations of pentanol and TBHQ, FTIR has been used [14–16]. FTIR spectroscopy is used for the observation of fuel stability which has been proposed as the most efficient and easiest method [17].

Table 1 Properties of fuel blends

Properties/samples	Density @ 15 °C	Kinematic viscosity @ 40 °C	Flash point	Pour point	Cloud point	Cetane index	Calorific value	Oxidation stability @ 110 °C
Units	kg/m³	cSt	°C	°C	°C	CCI	MJ/kg	h
Diesel	834	3.8	65	−4	6.2	53	43.32	–
B100	876	4.72	138	4.1	13.0	55.6	37.21	8.42
B100P1	876	4.72	138	4.1	13.0	55.6	37.21	23.80
B100P2	876	4.72	140	4.1	13.0	55.6	37.21	23.89
B100P3	877	4.73	141	4.1	13.3	55.8	37.26	26.82
B100P4	879	4.73	141	4.3	13.3	56.0	37.54	25.86
B100P5	879	4.84	143	4.3	13.3	56.2	37.62	25.40

2 Materials and Methodology

2.1 Rancimat Instrument

Figure 1 represents the Rancimat apparatus used for the study of antioxidant characteristics. By using 873 biodiesel Rancimat instrument, oxidation stability can be determined based on the EN14214 test method. The resistance to chemical and physical properties degradation by fuel is known as oxidation stability. The oxidation reaction can be controlled by the presence of hydro peroxides and peroxides [6]. For this test, the Indian standard IS15607 is used. From the Met Rohm manufacturer, the Rancimat instrument is obtained. For six hours of induction period, the biodiesel should be displayed at 110 °C as per the EN14214 standard for 6 h.

2.2 FTIR Spectroscopy

As shown in Fig. 2 PerkinElmer spectrum instrument has been used for FTIR analysis. For correlating the oxidation stability, the FTIR acts as an effective tool. This can be done by characterizing the specific molecular compounds with the absorption frequency range of 600–4000 cm^{-1}. For evaluating the fuel stability results, the C–H and O–H bonds were determined with the corresponding frequency range of 3000–3700 and 2800–3000 [14, 15].

Fig. 1 Metrohm 873 biodiesel Rancimat apparatus

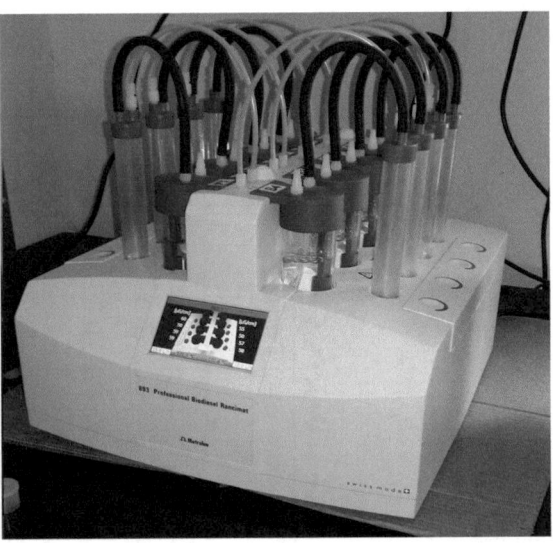

Fig. 2 PerkinElmer
spectrum instrument

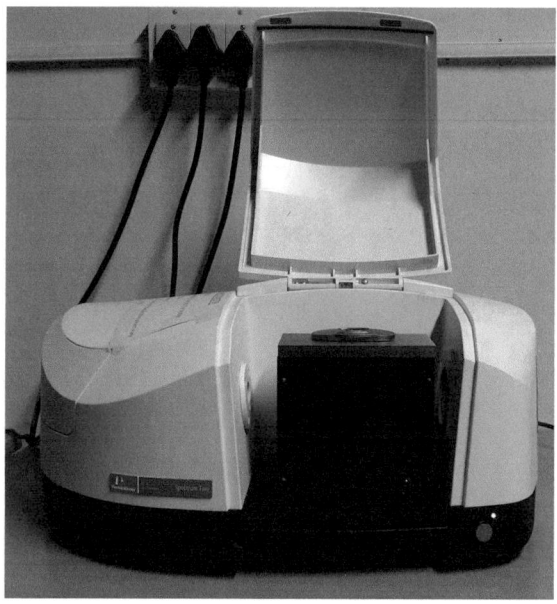

2.3 Thermogravimetric Analyzer

During combustion inside the engine cylinder biodiesel exhibits higher temperature which increases the NO_x emission [18, 19]. Figure 3 represents the thermogravimetric analyzer—TGA 2050 which is used in this paper to study the thermal stability. The main use of thermogravimetric analyzer is to analyze the resistance offered by fuel against degradation for thermal stability. When compared to higher temperature, the oxidation reaction is different at lower temperature. Diels–Alder reaction explains that at above 250 °C, it will initiate the oxidation reaction of fatty acid chains in a rapid manner [20]. As a result of elevated engine cylinder temperature, there is a chance of recirculation of fuel back into the fuel injection system which will lead to change in fuel physical and chemical properties [6]. The investigation of antioxidants can be obtained from 'onset temperature,' but for thermal stability there is no specific standard [21]. While investigating previous research papers, it has been observed that there has been no sufficient study concerning thermal stability of biodiesel [22].

2.4 Test Fuels

About 80% of biodiesel can be obtained from the Alexandrian laurel oil by the general process of transesterification [13, 23–25]. Reactants like high-grade methanol (20% v/v oil) and sodium hydroxide (0.9% w/w oil) were added with raw oil which has been leveled at 60 °C and at 300 rpm it is stirred. With the help of gravity separa-

Fig. 3 Thermogravimetric
analyzer—TGA

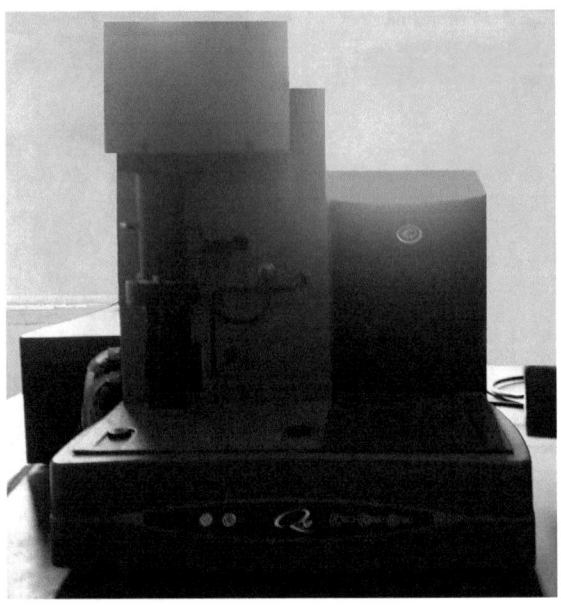

tion, the glycerol was removed after 12 h. Finally, C.I. biodiesel was obtained after
washing with distilled water [11]. After obtaining biodiesel, it has been dosed with
pyrogallol completely soluble in different concentrations ranging from 365, 650, 950,
1100, 1500 ppm. Table 1 represents the properties of test fuels which are determined
with the help of equipments in thermal laboratory, Government College of Technology, Coimbatore, India. The test fuel blends which are evaluated are as follows:
(a) pure biodiesel (B100), (b) 365 ppm pyrogallol (B100P1) + 100% biodiesel, (c)
650 ppm pyrogallol (B100P2) + 100% biodiesel, (d) 950 ppm pyrogallol (B100P3)
+ 100% biodiesel, (e) 1100 ppm pyrogallol (B100P4) + 100% biodiesel, (f) 1500 pp,
pyrogallol (B100P5) + 100% biodiesel.

3 Results and Discussion

3.1 Analysis of Thermal Stability

A sample quantity of 9 mg fuel was varied with oxygen and heated to 400 °C at
a rate of 10 °C/min. The process is performed to analyze the thermal stability of
biodiesel samples using onset temperature with the help of TGA 2050 thermogravimetric analyzer (TON). The fuel samples of B100P3 and B100P4 showed (68.65
and 50.78%) elevated TON when compared to B100. This was found out when the
biodiesel sample shows lower onset temperature when compared to biodiesel sample with antioxidant. Figure 4 shows higher onset temperature for all samples by the

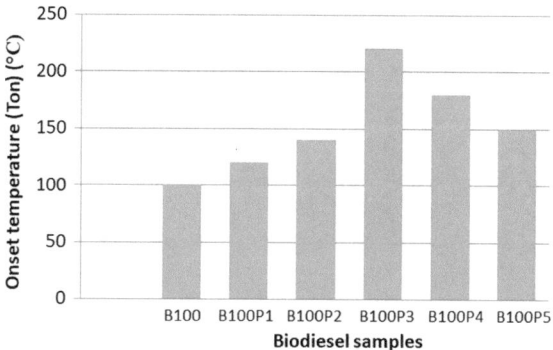

Fig. 4 Thermogravimetric analyzer onset temperature (T_{ON}) readings for all fuel blends

addition of pyrogallol antioxidant. Better thermal stability of sample is specified by elevated TON [6, 26]. Due to elevated temperature, the sample will lose its weight which will lead to the removal of secondary oxidation products.

3.2 Analysis of Oxidation Stability

The C.I. biodiesel shows higher stability of oxidation enhancement with the pyrogallol antioxidant. This can be observed from Table 1 in which the samples have been initially tested at 110 °C with the IP of above 7 h. From Table 1, it has been observed that the sample B100P3 shows higher induction period at 110 °C of about 26.82 h. With the help of automated Rancimat software, the obtained IP values are extrapolated at 20 °C, and also it has been estimated for years and hours in Table 2. Figure 5 shows the induction period difference between pure Alexandrian laurel and with the addition of pyrogallol antioxidant for the fuel samples by the following four temperatures such as 135, 140, 145, and 150 °C.

Up to 950 ppm (B100P3), the induction period increases with raise in pyrogallol antioxidant. When there is additional boost in induction period, it will lead to fall in the period level due to the breakage of chemical structure. Pyrogallol antioxi-

Table 2 Represents Rancimat measurements of all fuel blends at 30 °C

Samples	Extrapolated IP at 30 °C	
	(year)	(h)
B100	0.06	673.5
B100P1	0.51	4569.5
B100P2	0.64	5492.7
B100P3	1.2	12,468
B100P4	1.0	10,854
B100P5	0.89	7682.5

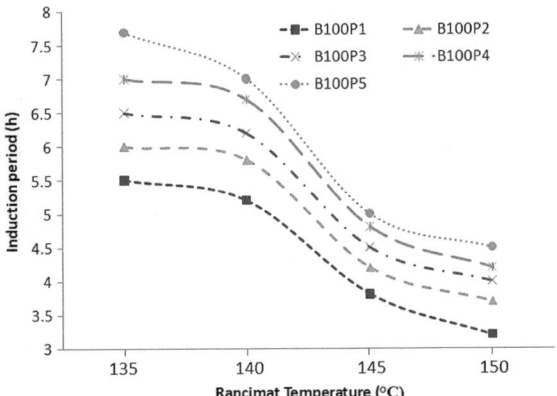

Fig. 5 Induction periods from Rancimat instrument for all fuel blends

dant involved in B100P3 sample resulted in 95.87% higher oxidation stability when compared to pure biodiesel. The breakage of hydrophilic and hydrophobic clusters resulted to diminish in the level of B100P4 and B100P5 oxidation stability. The above-mentioned characteristics can be clearly comprehensible with the help of Fig. 6 which has been obtained from FTIR spectrum.

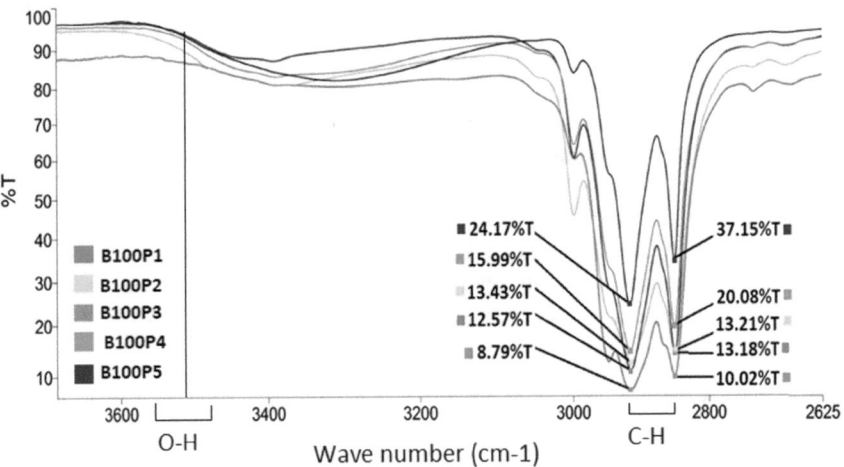

Fig. 6 Fuel samples spectrum from FTIR

3.3 Analysis of Storage Stability

Generally, the storage stability can be analyzed by the changes in acid value and kinematic viscosity of fuel blends. The acid value and kinematic viscosity changes over the period of 100 days for all fuel blends are shown in Figs. 7 and 8. From Figs. 7 and 8 for all the samples, it can be seen that with increase in the storage time, the acid value and kinematic viscosity increase. For this investigation, the KV and AV for all samples were dogged over an equal interval of 10 days in the aluminum containers at 30 °C. When there is an increase in kinematic viscosity level, there will be also raise in the level of acid value simultaneously. This is due to the oxidation reaction which leads to the formation of sediments and gums resulting in the formation of oxidized polymeric compounds and peroxides in the biodiesel samples [27]. Further during the period of storage, the formed hydro peroxides and oxides oxidized further in the acid value which has led to increase in the level of acid value [27]. By preventing the formation of peroxides, the boost in acid value and kinematic viscosity of fuel samples can be kept in control. This is possible by adding the pyrogallol antioxidant

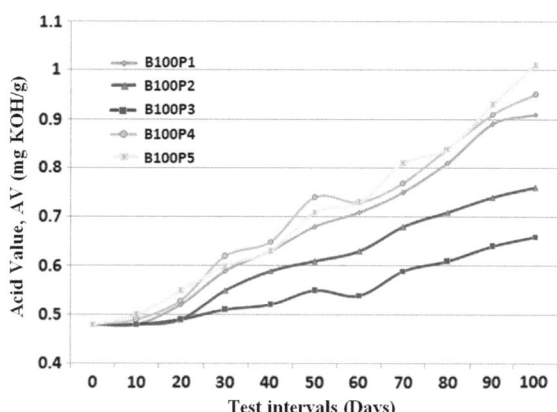

Fig. 7 Variation of acid value for all fuel samples stored for 100 days

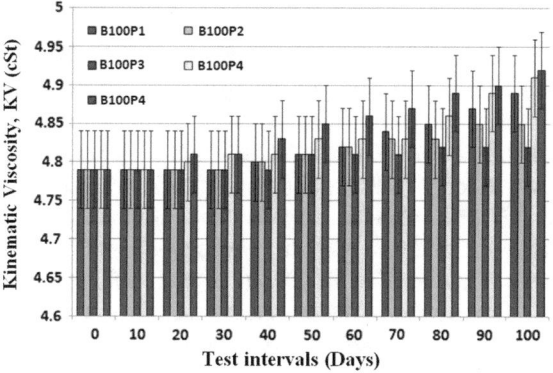

Fig. 8 Variation of kinematic viscosity for all fuel samples stored for 100 days

in the fuel samples which will show better storage stability as well as lessening the increase rate in acid value and kinematic viscosity. Both kinematic viscosity and acid value show a drastic raise in the level of pure biodiesel (B100) after 40 days of storage. Better resistance to increase in the level of acid value and kinematic viscosity is shown by pyrogallol antioxidant up to 950 ppm. This can be proved again from the FTIR spectrum graph which shows not a foremost change in the kinematic viscosity and acid value at 950 ppm (B100P3) for 150 days. Above 950 ppm due to the low-grade structure of B100P4 and B100P5 samples, the concentration level increases and the storage stability decreases.

3.4 Evaluation of Fourier-Transform Infrared Spectroscopy (FTIR)

The major use of FTIR spectroscopy is the collection of wide spectral range as well as infrared spectrum of emission or absorption of solid, liquid, or gas. With the help of specific molecular bonds which transmit the infrared spectrum regions, the graph has been plotted. For all samples, the FTIR spectrum data has been plotted as wave number (cm^{-1}) versus transmittance percentage (%) which is shown in Fig. 6. The chief components for the oxidation reaction are C–H and O–H bonds, respectively [11]. Lower presence of molecular compounds in the samples can be identified from the higher infrared transmittance in the FTIR graph. The most important factor to be considered in the biodiesel samples is to prevent the presence of large number of oxygen molecules which will lead to lower storage stability, thermal stability, and oxidation stability. Thus, the formation of hydro peroxides and peroxides levels in the biodiesel samples can be lowered which are also the main factors for higher oxidation reaction [11]. These factors can be clearly visualized from the FTIR graph in which the pure biodiesel (B100) shows that the strong O–H chain in $3521.2 \, cm^{-1}$ is the source for higher oxidation reaction. By analyzing with the pyrogallol additives, it has been observed that due to the fewer free radicals which involve less impact in the attraction of oxygen molecules, the results show an increment in oxidation stability, thermal stability, and storage stability than the previous obtained results. This will lead to delayed oxidation reaction in the antioxidant blends. Not only the above factors but also involves higher C–H bonds which have been obtained through the FTIR spectrum. The stronger C–H bonds of antioxidant surrounds the O–H bonds of biodiesel which will act as a cluster making it a very complex procedure to oxidize. Thus, the superior fuel stability is obtained by taking more time for breaking the cluster [11]. From Fig. 6 with the increase in pyrogallol concentration, B100P1 and B100P2 show increase in the C–H molecules with the display of 24.17%T and 15.99%T and also with the wave number of $2670.25 \, cm^{-1}$. It is also observed that raise in the level of B100P3 (i.e., up to 950 ppm) is due to the higher presence of C–H molecules in the sample. In the other terms from the above content, it can be clearly seen that up to 950 ppm the fuel stability in the B100P3 sample is 8.79%T at

2960.3 cm^{-1} higher when compared to B100P4 and B100P5. The lower fuel stability in B100P4 and B100P5 is observed due to increase in the pyrogallol antioxidant. The reason behind the degradation is due to the higher exposure hydrophilic entities to the oxygen molecules in the atmosphere as well as to the sample [11].

4 Conclusions

From the above investigation of pyrogallol antioxidant with the Alexandrian laurel biodiesel, the following conclusions were made for the thermal stability, oxidation stability, and storage stability.

(a) Using thermogravimetric analyzer (TGA), the thermal stability has been determined for all samples, and it has been observed that at 950 ppm the pyrogallol antioxidant shows 68.65% elevated T_{ON} (onset temperature) than the pure biodiesel.

(b) When compared to pure biodiesel, B100P3 shows about 95.87% higher induction period which has been obtained with the help of Rancimat instrument in which the oxidation stability was investigated.

(c) Due to the higher dispersion and diffusion of pyrogallol antioxidant with the biodiesel, the antioxidant volume concentration shows higher storage stability above 950 ppm which is referred to a second rate storage stability.

(d) The chemical structure of all fuel samples has been analyzed with the help of FTIR spectroscopy. From the FTIR spectroscopy, it has been observed that B100P3 shows higher C–H bond chains (8.79%T at 2960.3 cm^{-1}) which denotes that at higher rate the biodiesel will not get oxidized easily. In other hands for B100P4 and B100P5, there has been lower fuel stability which is due to deterioration of formed hydrophilic and hydrophobic clusters between molecular compounds.

References

1. Shameer, P. M., Ramesh, K., Sakthivel, R., & Purnachandran, R. (2017). Effects of fuel injection parameters on emission characteristics of diesel engines operating on various biodiesel: A review. *Renewable and Sustainable Energy Reviews, 67,* 1267–1281. https://doi.org/10.1016/j.rser.2016.09.117.
2. Shameer, P. M., Ramesh, K., Sakthivel, R., & Purnachandran, R. (2016). Assessment on the influence of compression ratio on the performance, emission and combustion characteristics of diesel engine fuelled with biodiesel. *Asian Journal of Research in Social Sciences and Humanities, 6*(12), 344–372. https://doi.org/10.5958/22497315.2016.01297.1.
3. BP Statistical Review of World Energy (2016, June). (65th edn). https://www.bp.com/content/dam/bp/pdf/energy-economics/statistical-review-2016/bp-statistical-review-of-world-energy-2016-full-report.pdf. Accessed Aug 3, 2017.

4. Shameer, P. M., & Ramesh, K. (2017). Experimental evaluation on performance, combustion behavior and influence of in-cylinder temperature on NO_x emission in a D.I. diesel engine using thermal imager for various alternate fuel blends. *Energy, 118,* 1334–1344. https://doi.org/10.1016/j.energy.2016.11.017.

5. Shameer, P. M., Ramesh, K., Purnachandran, R., & Sakthivel, R. (2017). Effects of injection timing and injection pressure on biodiesel fuelled engine performance characteristics: A review. *Asian Journal of Research in Social Sciences and Humanities, 7*(2), 310–330. https://doi.org/10.5958/2249-7315.2017.00093.4.

6. Pullen, J., & Saeed, K. (2012). An overview of biodiesel oxidation stability. *Renewable and Sustainable Energy Reviews, 16,* 5924–5950. https://doi.org/10.1016/j.rser.2012.06.024.

7. Yaakob, Z., Narayanan, B. N., Padikkaparambil, S., Unni, K. S., & Akbar, M. (2014). A review on the oxidation stability of biodiesel. *Renewable and Sustainable Energy Reviews, 35,* 136–153.

8. Pullen, J., & Saeed, K. (2012). An overview of biodiesel oxidation stability. *Renewable and Sustainable Energy Reviews, 16,* 5924–5950. https://doi.org/10.1016/j.rser.2012.06.024.

9. Yaakob, Z., Narayanan, B. N., Padikkaparambil, S., Unni, K. S., & Akbar, M. (2014). A review on the oxidation stability of biodiesel. *Renewable and Sustainable Energy Reviews, 35,* 136–153. https://doi.org/10.1016/j.rser.2014.03.055.

10. Atabani, A. E. & da S. Cesar, A. (2014). *Calophyllum inophyllum* L.—A prospective non-edible biodiesel feedstock. Study of biodiesel production, properties, fatty acid composition, blending and engine performance. *Renewable & Sustainable Energy Reviews, 37,* 644–655, http://dx.doi.org/10.1016/j.rser.2014.05.037.

11. Shameer, P. M., Ramesh, K., Sakthivel, R., & Purnachandran, R. (2017). Experimental evaluation on oxidation stability of biodiesel/diesel blends with alcohol addition by Rancimat instrument and FTIR spectroscopy. *Journal of Mechanical Science and Technology, 31* (1), 455–463. https://doi.org/10.1007/s12206-016-1248-5.

12. Shameer, P. M., Ramesh, K., Sakthivel, R., & Purnachandran, R. (2016). Assessment on the influence of compression ratio on the performance, emission and combustion characteristics of diesel engine fuelled with biodiesel. *Asian Journal of Research in Social Sciences and Humanities, 6*(12), 344–372. https://doi.org/10.5958/2249-7315.2016.01297.1.

13. Shameer, P. M., Ramesh, K., Purnachandran, R., & Sakthivel, R. (2017). Effects of injection timing and injection pressure on biodiesel fuelled engine performance characteristics: A review. *Asian Journal of Research in Social Sciences and Humanities, 7*(2), 310–330. https://doi.org/10.5958/2249-7315.2017.00093.4.

14. Shameer, P. M., Ramesh, K., Sakthivel, R., & Purnachandran, R. (2017). Experimental evaluation on oxidation stability of biodiesel/diesel blends with alcohol addition by Rancimat instrument and FTIR spectroscopy. *Journal of Mechanical Science and Technology, 31*(1), 455–463. https://doi.org/10.1007/s12206-016-1248-5.

15. Shameer, P. M., & Ramesh, K. (2017). FTIR evaluation on the fuel stability of *Calophyllum inophyllum* biodiesel: Influence of tertbutylhydroquinone (TBHQ) antioxidant. *Journal of Mechanical Science and Technology, 31*(7), 3611–3617. https://doi.org/10.1007/s12206-017-0648-5.

16. Saluja, R. K., Kumar, V., & Sham, R. (2016). Stability of biodiesel—A review. *Renewable and Sustainable Energy Reviews, 62,* 866–881. https://doi.org/10.1016/j.rser.2016.05.001.

17. Furlan, P. Y., Wetzel, P., Johnson, S., Wedin, J., & Och, A. (2010). Investigating the oxidation of biodiesel from used vegetable oil by FTIR spectroscopy: Used vegetable oil biodiesel oxidation study by FTIR. *Spectroscopy Letters, 43,* 580–585. https://doi.org/10.1080/00387010.2010.510708.

18. Shameer, P. M., & Ramesh, K. (2017). Experimental evaluation on performance, combustion behavior and influence of in cylinder temperature on NO_x emission in a D.I. diesel engine using thermal imager for various alternate fuel blends. *Energy, 118,* 1334–1344. http://dx.doi.org/10.1016/j.energy.2016.11.017.

19. Mohamed Shameer, P., & Ramesh, K. (2017). Study on clean technology-assisted combustion behavior and NO$_x$ emission using thermal imager for alternate fuel blends. *International Journal of Environmental Science and Technology.* https://doi.org/10.1007/s13762-017-1353-8.

20. Saluja, R. K., Kumar, V., & Sham, R. (2016). Stability of biodiesel—A review. *Renewable and Sustainable Energy Reviews, 62,* 866–881. https://doi.org/10.1016/j.rser.2016.05.001.

21. Nik, W. B. W., Ani, F. N., & Masjuki, H. H. (2005). Thermal stability evaluation of palm oil as energy transport media. *Energy Conversion and Management, 46,* 2198–2215. https://doi.org/10.1016/j.enconman.2004.10.008.

22. Jain, S., & Sharma, M. P. A. L. (2011). Thermal stability of biodiesel and its blends: A review. *Renewable and Sustainable Energy Reviews, 15,* 438–448. https://doi.org/10.1016/j.rser.2010.08.022.

23. Shameer, P. M., Ramesh, K., Sakthivel, R., & Purnachandran, R. (2017). Effects of fuel injection parameters on emission characteristics of diesel engines operating on various biodiesel: A review. *Renewable and Sustainable Energy Reviews, 67,* 1267–1281. https://doi.org/10.1016/j.rser.2016.09.117.

24. Shameer, P. M., Ramesh, K., Sakthivel, R., & Purnachandran, R. (2016). Assessment on the influence of compression ratio on the performance, emission and combustion characteristics of diesel engine fuelled with biodiesel. *Asian Journal of Research in Social Sciences and Humanities, 6*(12), 344–372. https://doi.org/10.5958/2249-7315.2016.01297.1.

25. Shameer, P. M., & Ramesh, K. (2017). Green technology and performance consequences of an eco-friendly substance on a 4-stroke diesel engine at standard injection timing and compression ratio. *Journal of Mechanical Science and Technology, 31*(3), 1497–1507. https://doi.org/10.1007/s12206-017-0249-3.

26. Nik, W. B. W., Ani, F. N., & Masjuki, H. H. (2005). Thermal stability evaluation of palm oil as energy transport media. *Energy Conversion and Management, 46,* 2198–2215. https://doi.org/10.1016/j.enconman.2004.10.008.

27. Karavalakis, G., Hilari, D., Givalou, L., Karonis, D., & Stournas, S. (2011). Storage stability and ageing effect of biodiesel blends treated with different antioxidants. *Energy, 36,* 369–374. https://doi.org/10.1016/j.energy.2010.10.029.

Artificial Intelligence Based Artificial Neural Network Model to Predict Performance and Emission Paradigm of a Compression Ignition Direct Injection Engine Under Diesel-Biodiesel Strategies

Kiran Kumar Billa, G. R. K. Sastry and Madhujit Deb

Abstract Environmental pollution as well as fast depleting fossil sources are key factors leading toward search for the alternative resources of energy. Unlike fossil fuels, biofuels are renewable source of energy. Biofuel is defined as any fuel whose vitality is obtained through a course of biological carbon change. Today, the use of biofuels has extended throughout the earth. Major producers include Asia, Europe and America and they consume too. Biodiesel is unpolluted burning alternative energy source from 100% renewable resources which can be termed as the fuel for the future. Many edible alternatives such as Palm oil, Sunflower, Soya Bean and, non-edible alternatives such as neem, canola, mahua, karanja, and jatropha are tested successfully with or without engine modifications. The results are encouraging. Biodiesel is a green fuel that can be mixed with petroleum diesel to produce biodiesel blends to run diesel machines. This action can reduce huge foreign exchange load on government, and helps in reducing harmful emissions. In the present study, *Pongamia pinnata* methyl ester with an oxygenative additive DEE is added with diesel. Further, with the help of artificial intelligence an attempt is made to find out the optimal blend. Artificial Neural Networks in MATLAB are used to serve the purpose. Furthermore, with the help of an ANN meta model an attempt is made to find out the optimal blend.

Keywords 2-EHN additive · NO_x · UHC · ANN · FFNN · MAPE · MSE

K. K. Billa (✉) · G. R. K. Sastry · M. Deb
Department of Mechanical Engineering, NIT, Agartala, Tripura, India
e-mail: billa2962@gmail.com

G. R. K. Sastry
e-mail: grksastry@gmail.com

M. Deb
e-mail: madhujit_deb@rediffmail.com

© Springer Nature Singapore Pte Ltd. 2020
S. K. Ghosh (ed.), *Energy Recovery Processes from Wastes*,
https://doi.org/10.1007/978-981-32-9228-4_7

1   Introduction

The swift depletion of petroleum energies and consistent price hikes of crude barrels have made a serious impact on the fuel and transport sectors and on the individual country's economy. Many alternatives were proposed by the researchers to analyze the characteristics of a diesel engines running on biodiesel derived from various vegetable oils, which may be congregated as edible and non-edible oils [1, 2] The investigations disclosed that vegetable oil esters provide improved performance and reduced emissions when compared to crude vegetable oils. Since the biodiesels are derived from plant and vegetable oils, they produce trifling greenhouse gas emissions [3]. The practice of using edible veggie oils such as like palm, sunflower, soya, rapeseed for fuel purposes may not be good idea as they are being used for cooking and can affect country's economy; i.e., it may fluctuate the cooking oil rates. Non edible variants should be used to avoid such an issue. Jatropha, rubber seed, *Pongamia pinnata* oil and linseed oil are examples of non-edible oils. *P. pinnata* is encouraged by the Indian government like a partial substitute to mineral diesel [4]. Proposed cultivation lands are side ways of railway tracks, dry lands of country side and government lands of hill sides to grow these shrubs. The Indian railways, one of the largest consumer of petroleum diesel started utilization of biodiesels to run some special engines. In addition, State governments also took initiative and they started using vehicles with bio fuels for transport and personal use. *P. pinnata* is easy to cultivate, grows faster and utilizes less water [5]. The by-products like seed crush can be used to feed cattle and even they can be used as a fertilizer in farm fields. The seed consists of viscous oil capable of having calorific value and is used to produce soap oil [6]. The practice of blending the *P. pinnata* methyl ester blends as fuels substitutions for diesel engines helps to minimize the energy demand and can minimize the huge foreign exchange to go out. The *P. pinnata* seed kernel has 46.5% viscous oil content [7]. The *P. pinnata* has palmitic (12.8%), stearic (7.3%), oleic (44.8%), linoleic (34.0%) acids in its chemistry [8].

1.1   *Motivation of the Present Study*

- To search an alternative fuel which ultimately improves the engine performance parameters as well as reduces the engine exhaust emission parameters by using *P. pinnata* oil methyl ester (PME).
- A single-cylinder four-stroke DICI engine with no engine modification is used to carry out the experiment with different blends of PME with fossil diesel.
- Proper blending with 2-EHN additive with the biodiesel-diesel fuel blends to enhance the performance of the biodiesel which in turn enhances the engine performance parameters and also reduces the engine exhaust emission parameters to make the fuel ideal.

- For optimization, Artificial Neutral Network approach is applied to predict the values of all required points within the ranges depending upon the values of input parameters coming out from experimental results.

At last, to establish the compatibility of *P. pinnata* oil biodiesel as a clean and environment friendly fuel for future use with distinguished effect in engine performance and exhaust emission.

In the present study, biodiesel is produced from non edible oil *P. pinnata*. The biodiesel is the methyl ester of *P. pinnata* oil. Its properties are similar to that of high-speed diesel. Before conducting the performance test of biodiesel in the Kirloskar made single-cylinder variable compression ratio engine, we need to design the experiment. In the present study, there are three input variables: load, compression ratio and blend of fuel. The blends can also be varied from 0 to 100% by any sort of variation as shown in the literature. As per the literature the load also can vary from no load to full load in any sort of variation. Moreover we choose the levels of these three input parameters or design factors as five as per the literature suggested. The load is varied from 20 to 100% with increment of 20%.

2 Pongamia Pinnata Methyl Ester

2.1 Production of Pongamia Pinnata Methyl Ester

It is estimated that *P. pinnata* oil has potential oils with a yearly production of 200 metric ton, from which only 6% is utilized commercially [6]. In most of the Asian villages these trees are visible and are the main production areas and village people make use of this tree products for various purposes. This investigation highlights the efforts to produce biodiesel from *P. pinnata* from rural Indian forests and the experiments in the laboratory meticulously approves with the testified literature that the attendance of high Free Fatty Acid marks transesterification reaction hard because of the development of soap with alkaline catalyst. In the present virtue of experiments, the another route of using acid catalyst was implemented for the biodiesel production from *P. pinnata* oil [7–9]. Typically *P. pinnata* oil, collected for the present study, was detected to comprise 3.2% of FFA.

2.2 Transesterification Process

Transesterification or sometimes alcoholysis is the dislocation of alcohol group from an ester by another in a manner similar to hydrolysis, with the exception of alcohol is utilized instead of water. This practice has been widely utilized to reduce the high viscous triglycerides.

The transesterification chemical reaction is epitomized by the following general equation:

$$RCOOR' + R''OH \longleftrightarrow RCOOR'' + R'OH$$

If methyl alcohol is used in this course it is termed as methanolysis and its glyceride is represented. Transesterification is a reversible reaction. However, the presence of a catalyst (a strong acid) accelerates the conversion.

3   Experimental Setup and Procedure

3.1   Preparation of Modified Fuels

The test fuels used for the present investigation are biodiesel blends, derived from *P. pinnata*. The kinematic viscosity, specific gravity, and Calorific value of the bio diesel were measured by means of standard equipment and are 5.2 cSt at 32 °C, 0.93660 and 34.5 MJ/kg, respectively. The fuel additive used in this investigation is Di Ethyl Ether (DEE), density of 7.13 g/mL. The dosing level of the DEE (by volume) in the base fuel was 2%.

3.2   Determination of Fuel Properties

The kinematic viscosity, flash point, fire point, pour point and cloud point were measured by means of standard test methods. The viscosity was gauged by means of the Redwood viscometer [10]. A Cleveland apparatus for flash point and fire point [11, 12].

3.3   Details of Testing Equipment

A single-cylinder, water-cooled four-stroke DICI engine was used to investigate the performance and emission profiles shown in Fig. 1. The performance test is carried out on a single cylinder variable compression ratio DI diesel engine using high-speed diesel, methyl ester of *P. pinnata* oil and their blends with diesel. The engine is assembled and coupled with an eddy current dynamometer. The arrangement of experimental setup used for carrying out the present study is shown in Table 1. The load range taken is from 3 to 12 kg.

During experiment, fuel consumption is measured by a burette and a stop watch, the engine exhaust (CO, HC, CO_2, O_2 and NO_X) is analyzed and calculated by AVL

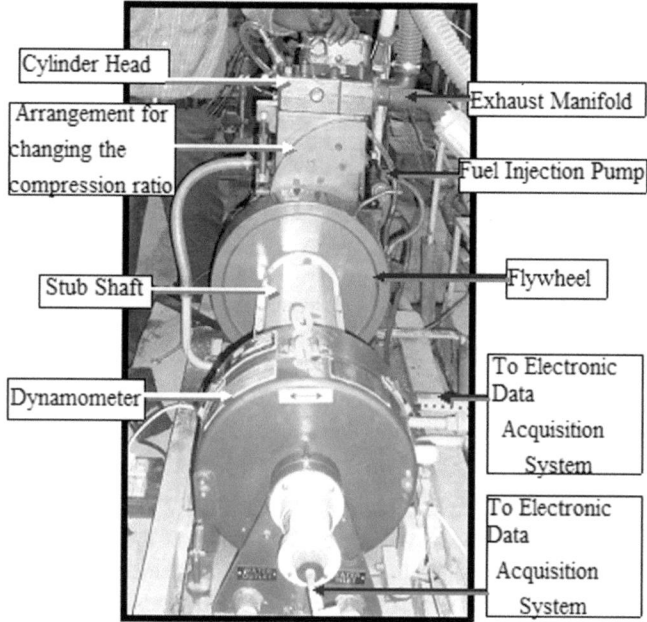

Fig. 1 Schematic diagram of experimental setup

Table 1 Test rig details

Make	Kirloskar
Model details	TV 1 (VCR)
Type	DI-type 4-stroke, water-cooled
Cylinder	One
Rated power	3.5 kW @ 1500 RPM
CR	12:1–18:1
Bore	87.50 mm
Stroke	110.0 mm
Injection timing	23° bTDC
Loading type	Eddy current dynamometer

DIGAS 444 gas analyzer fitted with DIGAS SAMPLER at the exhaust. Table 1 shows the engine details.

Table 2 Properties of the sample fuels

Property	Pong. pinn.	B100	B10	B20	B30	Diesel
Density (kg/m³)	912	898	856	862	868	850
Pour point (°C)	−4	4	3.5	3.6	3.7	3.1
Cloud point (°C)	14.6	10.2	7	7.1	8.2	6.5
Fire point (°C)	244	199	91	93	96	78
Flash point (°C)	242	196	89	91	95	76
Cetane number	38.0	57.9	48.2	47.9	47.6	49
Calorific value (MJ/kg)	34.5	39.22	43.32	42.23	41.34	44.82
Kinematic viscosity (cSt)	27.79	5.46	3.23	3.49	3.77	4.842

3.4 Methodology

The fuels used in this study are standard diesel and PME (99.9% purity, Laboratory used). The blending was done on volume basis as we know that biodiesel is miscible diesel in all proportions. Hence there is no problem of miscibility of *P. pinnata* biodiesel with diesel. For this experiment we have used blends of diesel and biodiesel in following proportion. They are mentioned below (Table 2):

1. D100—sample containing 100% diesel fuel.
2. B10—Sample containing 10% biodiesel + 0.5% EHN + 89.5% diesel.
3. B20—Sample containing 20% biodiesel + 0.5%EHN + 79.5% diesel.
4. B30—Sample containing 30% biodiesel + 0.5% EHN + 69.5% diesel.

4 Results and Discussion

4.1 Load Versus BP Graph

Figure 2 shows the disparity of brake power with regard to load for diesel fuel and *P. pinnata* biodiesel and additives. It can be observed from the figure that there is no significant change in brake power. All blends, i.e., B10, B20, B30 including diesel gave more or less the same readings. In specific 6% loss in Brake power is observed, this is because of the low energy content of *P. pinnata* blends.

4.2 Load Versus BSFC Graph

Figure 3 shows the disparity of brake specific fuel consumption with respect to load for diesel fuel and *P. pinnata* biodiesel and additives. It can be observed from the

Fig. 2 Load versus BP graph

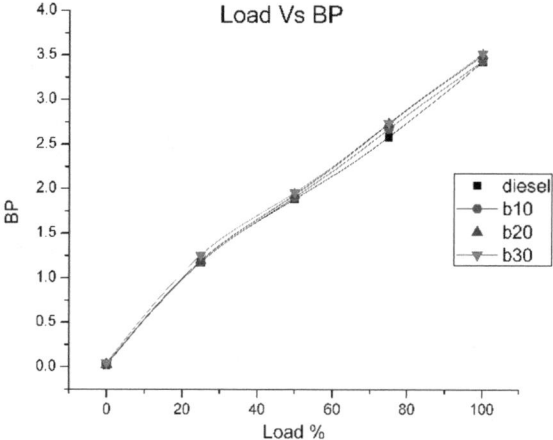

Fig. 3 Load versus BSFC graph

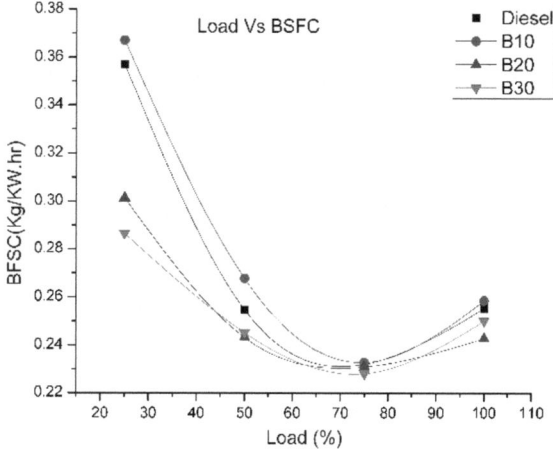

figure that shows b20 and b30 blends shows lower specific fuel consumption when compared to the conventional diesel But b10 biodiesel blend shows higher specific fuel consumption at any load. so should prefer b10 when for lower fuel consumption.

4.3 Load Versus BSEC Graph

Figure 4 shows the disparity of brake specific energy consumption with respect to load for diesel fuel and *P. pinnata* biodiesel and additives. It can be observed from the figure that shows B20 and B30 blends shows lower specific energy consumption when compared to the conventional diesel but B10 biodiesel blend shows higher specific energy consumption at low load. But after half load energy consumption decreases.

Fig. 4 Load versus BSEC graph

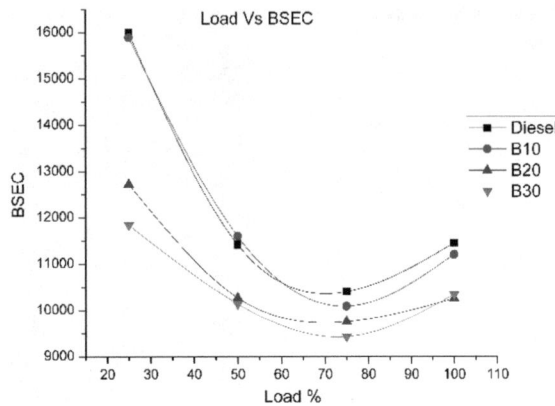

4.4 Load Versus BTE Graph

Figure 5 shows the disparity of CO with regard to load for diesel fuel and *P. pinnata* biodiesel and additives. It can be seen that the BTE of B20 is very close to diesel at higher loads. However all biodiesel samples were performing satisfactorily when compared to diesel. This is because of the excess oxygen portion in the biodiesel and the oxygenative additive helped in complete combustion and maximum utilization of the heat content in the biodiesel samples.

Fig. 5 Load versus BTE graph

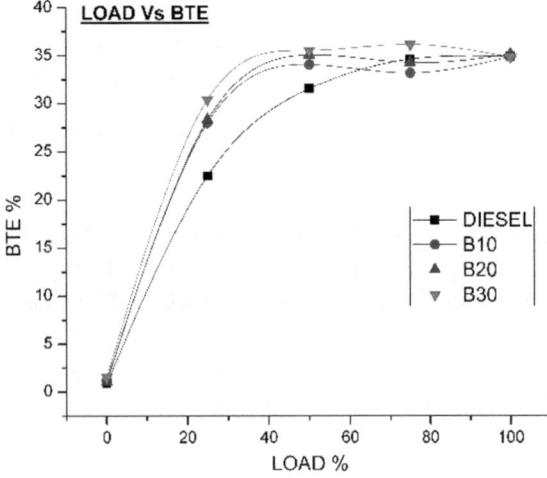

Fig. 6 Load versus carbon monoxide graph

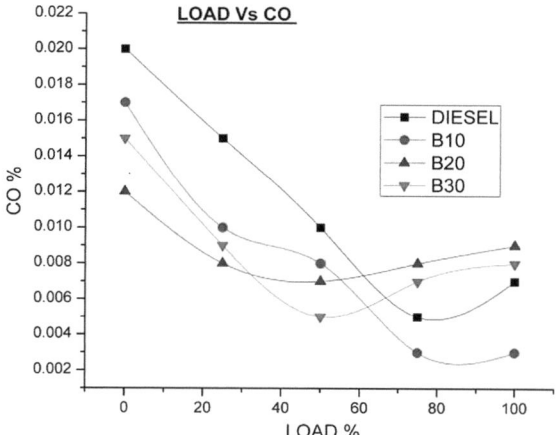

4.5 Load Versus CO Graph

Figure 6 shows the disparity of CO with regard to load for diesel fuel and *P. pinnata* biodiesel and additives. It is one of the prime objectives of the project. CO emissions are very less up to half load for any *P. pinnata* blend but after half load the CO emissions of B20 and B30 slowly creeped up. This is because of excess oxygen content in the *P. pinnata* blends which resulted in complete combustion and formed carbon dioxide instead of carbon monoxide.

4.6 Load Versus Carbon Dioxide Graph

Figure 7 shows the disparity carbon dioxide with regard to load for diesel fuel and *P. pinnata* biodiesel and additives. It is clearly seen that from graph all the blend shows low emissions of carbon dioxide at low loads and slightly increases at half loads which is negligible and again decreases at full loads when compared with conventional diesel.

4.7 Load Versus Hydro Carbons Graph

Figure 8 shows the disparity hydrocarbons with regard to load for diesel fuel and *P. pinnata* biodiesel and additives. Due to the excess oxygen present in the biodiesel and the oxygenative additive DEE complete combustion took place and formation of HC gone down. It is clear that B30 showed least HC production from the graph.

Fig. 7 Load versus carbon
dioxide graph

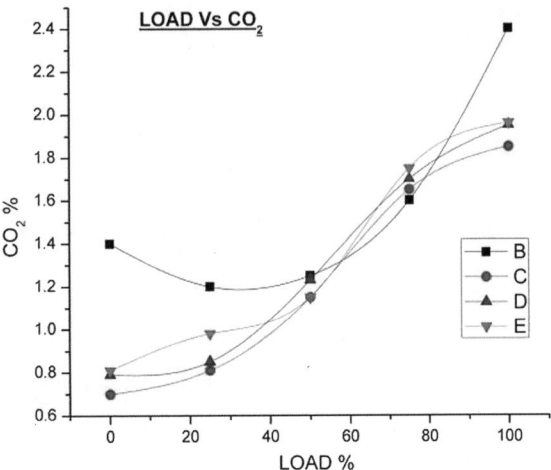

Fig. 8 Load versus HC
graph

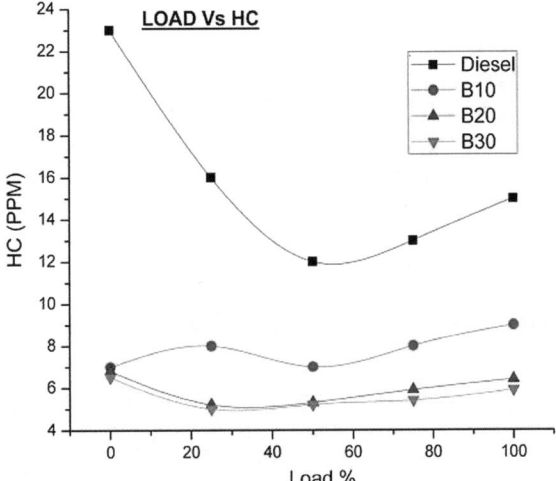

4.8 Load Versus Exhaust Gas Temperature Graph

Figure 9 shows the disparity of exhaust gas temperature with regard to load for
diesel fuel and *P. pinnata* biodiesel and additives. It is a degree of performance as
progressive exhaust gas temperatures results in greater heat release in combustion
chamber and hence giving probability to rise brake thermal efficiency. From the
graph it is clearly seen that that EGT is high in all the blends when compared to
conventional diesel.

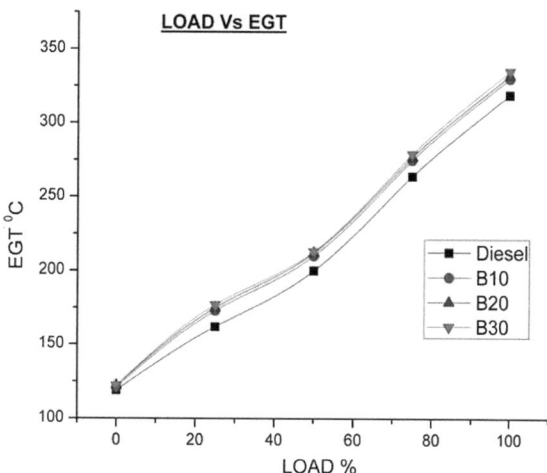

Fig. 9 Load versus EGT graph

4.9   Load Versus NO_x Graph

Figure 10 shows the variation of NO_x with respect to load for diesel fuel and *P. pinnata* biodiesel and additives. On the account of the excess oxygen present in the biodiesel complete combustion took place and temperature inside the combustion chamber increased thereby increasing the NO_x emission. From the graph it is clear that B30 showed maximum NO_x emissions. It cannot be completely stopped but it can be minimized like EGR techniques.

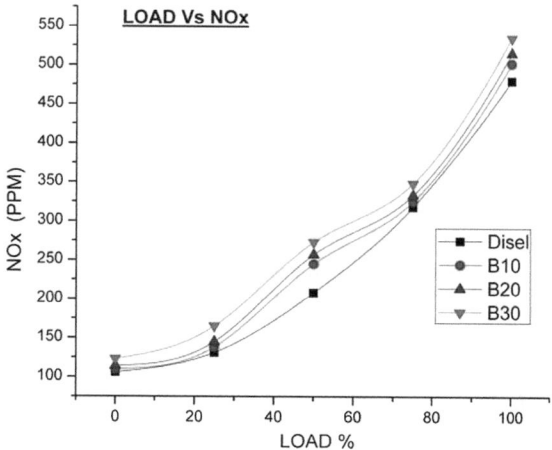

Fig. 10 Load versus NO_x graph

5   Artificial Nueral Networks (ANN)

ANN technique is a beneficial mathematical tool that has been adopted to predict responses required extended (experimentally) time and sophisticated devices such as internal combustion engines. Essentially, ANN denotes to a computation configuration method exhibited on biological processes, primarily on the implementation of human brain, involving number of interconnected meting out elements termed as neurons, which process data based on their active state with reference to inputs. Figures 11 and 12 demonstrates the typical architecture and the modelling tree of the proposed neural network model.

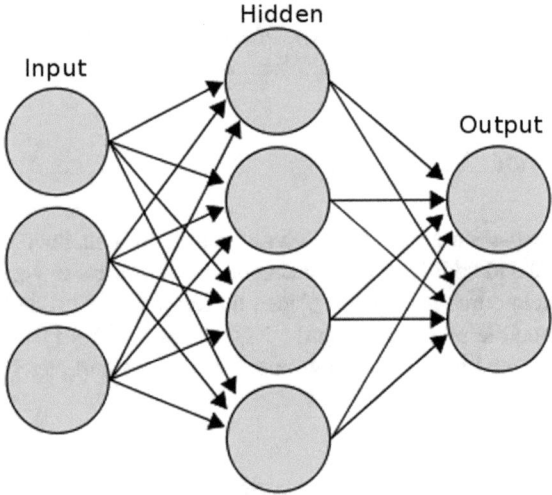

Fig. 11 Typical neural network model

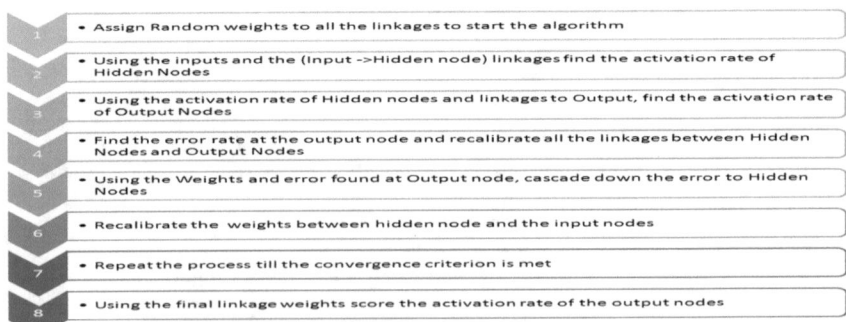

Fig. 12 ANN modeling tree

5.1 Development of ANN Model

Normalization of data which bring homogeneity among the engine responses is done by using the following formula

$$\frac{\text{Actual value}}{\text{Maximum} - \text{Minimun}} \times (\text{High} - \text{Low}) + \text{Low}$$

In the present research feed-forward neural network with back propagation technique is being employed. ANN has been cast-off considering innumerable model configurations to attain the optimum engine routine and exhaust gas discharge prediction. The developed ANN models is being trained using the collected experimental paradigm. Performance of ANN is subject to the data presented to it, hence mounting of input and output information is vital. In the current investigation, an ANN modeling is being used to predict the relationship of BSEC, BTE, NO_x, EGT, UHC and CO with load percent and percentage of biodiesel as input parameters. The relationship between the preferred input variables and anticipated outputs in the comprehension of the current study which has been studied in the association matrix bear proof to the inconsistent influences of the input parameters on the anticipated emission and performance responses. From the dataset, about a total of 20 samples are used in this model, 60% of the data (12 samples) dataset has been arbitrarily assigned as the training, while the 20% of dataset (4 samples) are utilized for testing and the residual 20% is used to estimate and validation. The developed Model is run using MATLAB toolbox. Multi-layer perception network (MLP) is used in nonlinear mapping among the inputs and the output paradigm. To an additional second derivative of error data and automatic internal alterations that are prepared to the learning constraints. In the present algorithm, the network weights as well as biases are initialized arbitrarily at the commencement of the training chapter. Error minimization method is achieved by means of a gradient descent rule. There are two input and six output parameters in the experimental test. The two input variables are considered to be as 'load percent' expressed in percentage and the 'Biodiesel percentage' also in terms of percentage. The six outputs have been outlined as BSEC, BTE, NO_x, EGT, UHC and CO on MATLAB toolbox, which was also selected by way of our iteration solver. The developed ANN model is limited to the tentative data which is obtained on the existing engine setup. The ANN edifice has established with double hidden layer and 16 neurons evidenced to be the best ANN selection in this replication subsequently the R^2 values are observed mostly closer to unity. During the constantly ANN model training, the minimum gradient of 10^{-5} and 1000 epochs are used as ending criteria. Trained model is simulated for all input parameters to comprehend corresponding individual output responses of the model. By means of targets as well as outputs of the model, regression coefficients, MAPE and MSE have been estimated using the following expressions [13]

$$\text{Regression Coefficient}(R) = \sqrt{1 - \left(\frac{\sum_{i=1}^{n}(e_i - p_i)^2}{\sum_{i=1}^{n} p_i^2}\right)}$$

where n is the total number of datasets, e_i is the experimental value and p_i is the network predicted value. The R-value was set to be limited to 25 neurons in the hidden layer. Thus, the network with 15 neurons in the hidden layer was found satisfactory. To attain an accurate outcome, a regression scrutiny of outputs and anticipated targets is performed and is presented. The precision of the training process is validated by means of the testing method.

The testing course involves in presenting a totally unique set of experimental dataset to test the precision of the trained network in forecasting the specific engine responses. A broad measure of the precision of the testing segment is the MSE shown in following equation.

$$\text{Mean Square Error} = \frac{1}{n}\left\{\sum_{i=1}^{n}(e_i - p_i)^2\right\}$$

The mean square error (MSE) and correlation coefficient (R^2), displayed in the table, are also computed when matched with the anticipated and investigational values. MSE provides data on the temporary performance of the model by authorizing term-by-term evaluation of actual deviation between the predicted and measured values. The MSE has been always positive and a zero value to be ideal. MSE delivers a pointer of the prognostic error relative to the precise value. Lower the MSE demonstrates a superior correlation among the prophesied and experimental outcomes. The R^2 value offers an another pointer between the prophesied and experimental information, Where R^2 values nearby to 1 signifies the most precise prediction.

5.2   Results and Discussions

ANN model training was utilized for number of times with 1000 iterations. Levenberg–Marquardt training procedure through logarithmic sigmoid transfer function employed for layer-1 as well as for layer-2 respectively, produced paramount regression, MAPE and MSE. It was observed that MSE is greater for smaller number of neurons, reaffirming the point that the smaller number of neurons determines decision-making vigorous. Although, MSE is greater for greater number of neurons as tuning of weights to diminish error is time-consuming. The optimum figure of neurons why ever MSE was perceived to be insignificant is 16. In acquirement of constraints to access the propinquity of actual and anticipated values, MSE alone was not adequate. Pursuant to the case of training algorithm, even though MSE is in the tolerable range, MAPE and regressions may not in the acceptable limits. Thus the skillful ANN model was developed by enchanting regression, MSE and MAPE in place of a valuation standard.

$$\text{Mean Absolute Percentage Error(MAPE)} = \left\{ \frac{100}{n} \sum_{i=1}^{n} \left| \frac{(e_i - p_i)}{(e)} \right| \right\} \%$$

Levenberg–Marquardt training function through logarithmic sigmoid transfer function, the ANN prophecies for proposed test cases demonstrated outstanding overall agreement catalogues with the experimental responses where it attained a 99.47% regression. Statistical studies for Brake Specific Energy Consumption (BSEC) has shown that the proposed ANN model gave least MSE content of 0.0141 [14]. The assurance of the ANN prophesied BTE is evident in the tenacious concurrency in association with the experimental data for all iterations of experimental data as evident from Fig. 13a–g. It is also witnessed that the ANN affords decent level of accuracy in demonstrating the engine responses. This has been supported by the reliable concurrency of the ANN anticipated values with the experimental dataset for the total series of observations. Likewise, NO_x emissions also anticipated precisely as obvious from the similarity with the observed data. From the tentative results we conclude that there is a substantial reduction in the CO, CO_2 and HC emission profiles because of superior combustion characteristics demonstrated by the pilot test fuels. Table 3 shows Regression values at various neurons and the maximum regression is achieved at a topology of 16 neurons as shown in Fig. 14.

6 Conclusions

NO_x levels were a bit higher than the petro diesel due to the reason that excess oxygen levels in the pilot fuel sample. We can witness that the emission profiles predicted with FFNN that track a certain trend have a poorer error percent than matched to those that doesn't track a certain flow. Additionally, the emissions that disagree approximately in a linear manner have a nominal percent of error than those that track a quadratic or a cubic equation. Therefore contingent to the nature as well as the tendency of emission profile we can practice FFNN to forecast the emission profiles depending upon the obligatory level of precision and thus providing the required emission profiles without resonant the actual experimental investigation. *P. pinnata* methyl ester seems to have a prospective to practice as alternative fuel in DICI engines deprived of any modification. Based on the above results the conclusions were drawn on whole as laid below.

1. Blending diesel declines the viscosity substantially.
2. It is also concluded from the experiment that practice of adding additives with diesel and biodiesel blends has increased the cetane number, lubricity, and stability of the testing fuel which resulted into improved performance with the PME blends.

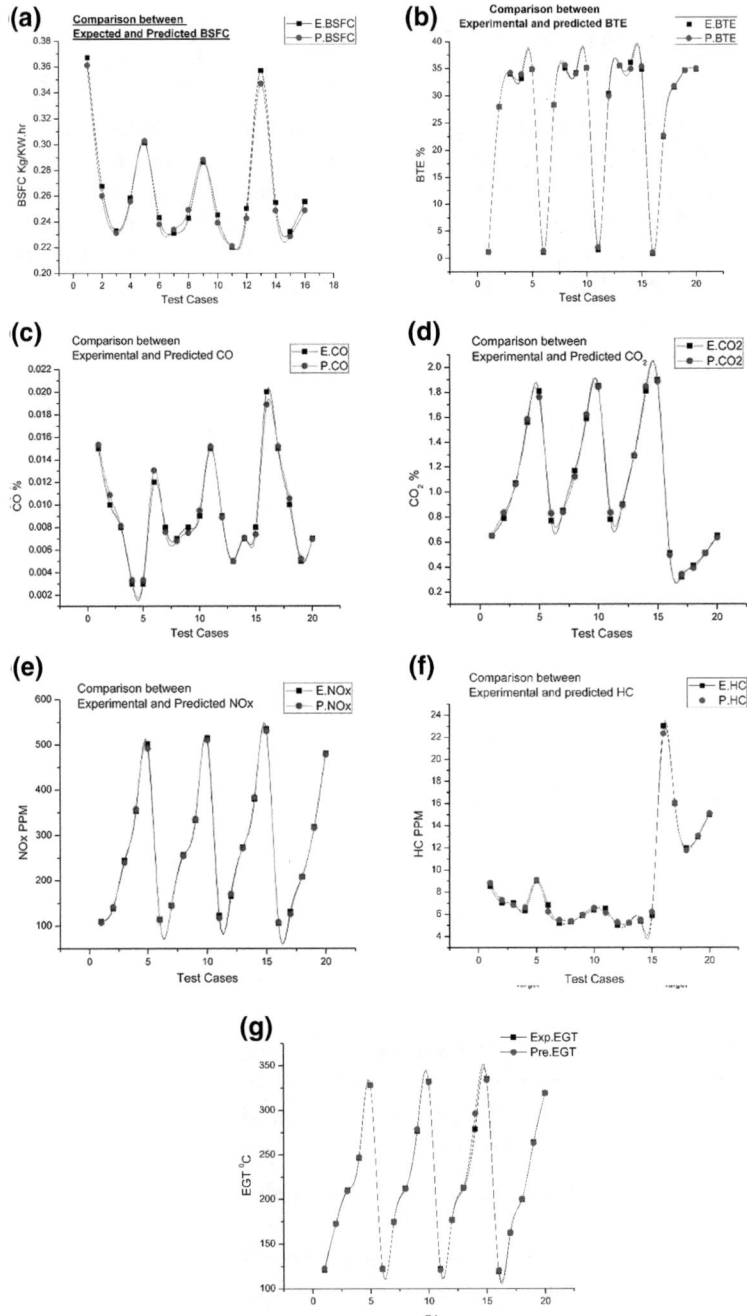

Fig. 13 a–g Comparison of experimental versus predicted properties

Table 3 Topology of various neurons

Neurons	Regression coefficient (r)			
	Training	Validation	Testing	Overall
10	0.96038	0.99807	0.99824	0.97078
11	0.98648	0.95858	0.99873	0.98466
12	0.98958	0.9592	0.8598	0.96496
13	0.97214	0.97223	0.97846	0.97329
14	0.99736	0.96716	0.97593	0.98805
15	0.97133	0.99875	0.99883	0.97799
16	**0.99596**	**0.99225**	**0.99211**	**0.9947**
17	0.96728	0.99077	0.99865	0.97372
18	0.98602	0.9679	0.84972	0.96468
19	0.9726	0.98899	0.99488	0.97844
20	0.92631	0.99713	0.99727	0.95181
21	0.97295	0.88313	0.99756	0.96326
22	0.95315	0.99412	0.99301	0.96728
23	0.94956	0.98571	0.99682	0.96064
24	0.94375	0.97692	0.97248	0.95204
25	0.97161	0.95927	0.94135	0.96417

3. It was instituted by the experiment that the blends of PME, DEE and diesel could be practiced successfully with satisfactory performance than pure diesel capable to a certain limit.
4. From experiment it is concluded that B20 could replace the diesel for diesel engine for attainment of better performance.
5. The brake thermal efficiency was slightly better than pure diesel fuel.
6. Brake specific fuel consumption is lesser for PME blends than diesel at all the load conditions.
7. The exhaust gas temperature is instituted to increase with percentage of PME in test fuels in accordance with coarse spray foundation and delayed combustion.
8. ANN is found to be capable of predicting the engine output responses with regression values close to the unity and MAPE and MSE values under acceptable threshold.

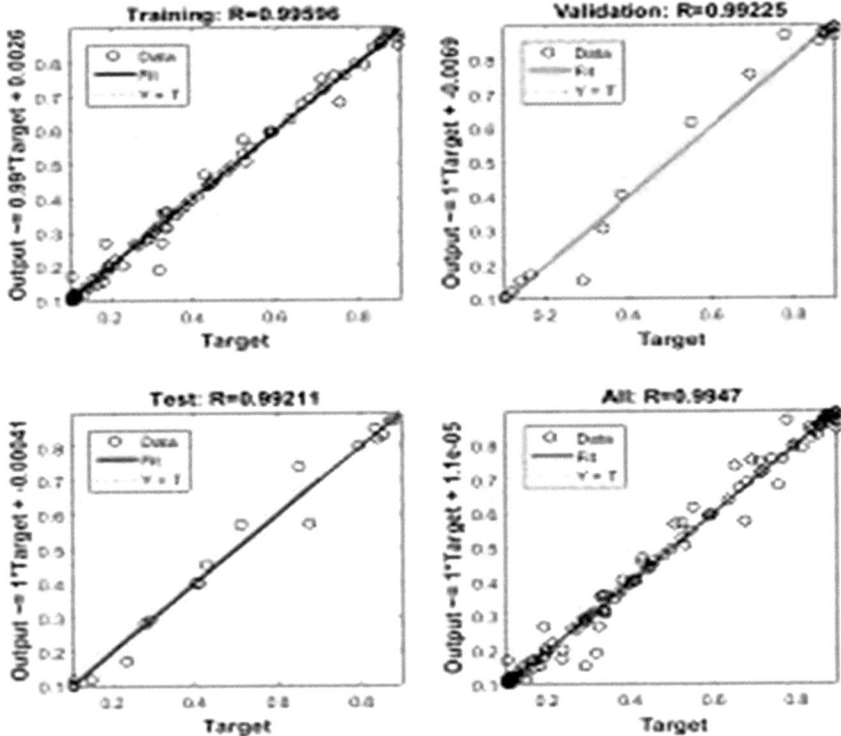

Fig. 14 Best topology

References

1. Raheman, H., & Phadatare, A. G. (2004). Biomass & bioenergy, emissions and performance of diesel engine from blends of karanja methyl ester (biodiesel) and diesel. *Energy, 27,* 393–397.
2. Puhan, S., Vedaraman, N., Ram, B. V. B., Sankarnarayanan, G., & Jeychandran, K. (2005). Mahua oil (*Madhuka indica* seed oil) methyl ester as biodiesel preparation and emission characteristics. *Biomass and Bioenergy, 28,* 87–93.
3. Aziz, A. A., Said, M. F., & Awang, M. A. (2005). *Performance of palm oil based biodiesel fuels in a single cylinder direct injection engine.* Malaysian Palm Oil Board Report.
4. Avinash Kumar Agarwal. (2007). Biofuels (alcohols & biodiesel) applications as fuel for internal combustion engine. *Progress in Energy and Combustion Science, 33,* 233–271.
5. Raheman, H., & Ghadge, S. V. (2007). Performance of compression ignition engine with mahua biodiesel. *Fuel, 86,* 2568–2573.
6. Ghosal, M. K., Das, D. K., Prada, S. C., & Sahoo, N. (2008, October). Performance study of diesel engine by using mahua methyl ester (biodiesel) and its blends with diesel fuel. *Agricultural Engineering International, 10.*
7. Hanumantha Rao, Y. V., Sudheer Voleti, R., Hariharan, V. S., & Sitarama Raju, A. V., Jatropha oil methyl ester and its blends used as an alternative fuel in diesel engine.
8. Kalbande, S. R., & Vikhe, S. D. (2008, February). Jatropha and karanja biofuel: An alternative fuel for diesel engine. *ARPN Journal of Egg & Applied Sciences, 3,* 7–13.

9. Bijou, B., Naik, M. K., & Das, L. M. (2009). A comparative evaluation of compression ignition engine characteristics using methyl & ethyl esters of karanja oil. *Renewable Energy, 34,* 1616–1621.
10. *Karanja—A potential source of biodiesel.* A report by National Oilseeds and Vegetable Oils Development Board, Government of India, Ministry of Agriculture (2008).
11. Sahoo, P. K., Das, L. M., Babu, M. K. G., Arora, P., Singh, V. P., Kumar, N. R., et al. (2009). Comparative evaluation of performance and emission characteristics of atrophy, karanja and polanga based biodiesel as fuel in a tractor engine. *Fuel, 88,* 1698–1707.
12. Murugesan, A., Marana, C., Subramanian, R., & Nedunchezhian, N. (2009). Biodiesel as an alternative fuel for diesel engines—A review. *Renewable and Sustainable Energy Reviews, 13,* 653–662.
13. Prasada Rao, K., Victor Babu, T., Anuradha, G., & Appa Rao, B. V. (2016). IDI diesel engine performance and exhaust emission analysis using biodiesel with an artificial neural network (ANN). *Egyptian Journal of Petroleum.*
14. Ismail, H. M., Ng, H. K., Queck, C. W., & Gan, S. (2012). Artificial neural networks modelling of engine-out responses for a light-duty diesel engine fuelled with biodiesel blends. *Applied Energy, 92,* 769–777
15. Kandasamy, M. M. K., & Thangavelu, M. (2004, December). Operational characteristics of diesel engine run by ester of sunflower oil and compare with diesel fuel operation. In: *International Conference, Sustainable Energy and Environment.*
16. Devan, P. K., & Mahalashmi, N. V. (2009). Utilization of unattended methyl ester of paradise oil as fuel in diesel engine. *Fuel.*
17. Nurun Nabi, Md., Mustafizur Rahman, Md., & Shamim Akhter, Md. (2009). Biodiesel from cotton seed oil and its effect on engine performance and exhaust emissions. *Applied Thermal Engineering, 29,* 2265–2270.
18. Channapattana, S. V., Pawar, A. A., & Kamble, P. G. (2017). Optimisation of operating parameters of DI-CI engine fueled with second generation Bio-fuel and development of ANN based prediction model. *Applied Energy, 187,* 84–95
19. Billa K. K., Sastry G. R. K., & Deb M. (2018). A Novel Comparison of Two Artificial Intelligent models for estimating the Kinematic Viscosity and Density of Cottonseed Methyl Ester. *International Journal of Computational Intelligence & IoT.*

Comparative Analysis of Experimental and Simulated Performance and Emissions of Compression Ignition Engine Using Biodiesel Blends

M. U. Kaisan, S. Abubakar, S. Umaru, B. Dhinesh, P. Mohamed Shameer, K. Sekar, P. Mohamed Nishath and J. Senophiyah Mary

Abstract The physicochemical properties of different biodiesel feedstock from past literature were reviewed, and their mean and standard deviation values calculated. The performance of biodiesel from cotton, jatropha and neem in a stationary multi-cylinder Compression Ignition engine at full load with variable speeds of 1000, 1500, 2000 and 2500 rpm were investigated and the composition of their exhaust gas emissions were recorded and analyzed. A model was developed for the same compression ignition engine with the test rig's specifications using the GT-Power modeling software. The properties of cotton, jatropha and neem obtained from the experimental works were used to validate the modeled engine during simulation, while the performance of biodiesels from the secondary data sources was further

M. U. Kaisan (✉) · S. Abubakar · S. Umaru
Department of Mechanical Engineering, Ahmadu Bello University, Zaria, Nigeria
e-mail: mukaisan@abu.edu.ng

S. Abubakar
e-mail: abubakarshitu88@gmail.com

S. Umaru
e-mail: bnumar@yahoo.com

B. Dhinesh
Department of Mechanical Engineering, Mepco Schlenk Engineering College, Sivakasi, Virudhunagar, Tamil Nadu, India
e-mail: dhineshbala91@mepcoeng.ac.in

P. Mohamed Shameer
Department of Mechanical Engineering, V. V. College of Engineering, Tirunelveli 627657, Tamil Nadu, India
e-mail: pmohamedshameer@gmail.com

K. Sekar
Department of Mechanical Engineering, National Institute of Technology, Calicut, Kerala, India
e-mail: sekar@nitc.ac.in

P. Mohamed Nishath · J. Senophiyah Mary
Government College of Technology, Coimbatore 641013, India
e-mail: pmdnishath@gmail.com

J. Senophiyah Mary
e-mail: senophiyah.mary8@gmail.com

© Springer Nature Singapore Pte Ltd. 2020
S. K. Ghosh (ed.), *Energy Recovery Processes from Wastes*,
https://doi.org/10.1007/978-981-32-9228-4_8

simulated under same conditions. The engine performance simulation in GT-Power confirmed that the software is valid to be used in compression ignition engine simulation with biodiesel from any feedstock and at all speeds because both experimented and simulated results are very close with identical peaks. Pure biodiesel samples produced the lower CO, NO_x, CO_2 and SO_2 emissions than diesel.

Keywords Biodiesel · GT-Power · Simulation

1 Introduction

Biodiesel is an alternative diesel fuel which is produced by trans-esterification of plants oils and animal fats [11, 11–14, 18, 19, 22]. Biodiesel is chemically defined as the monoalkyl esters of long-chain fatty acids resulting from sustainable biological feedstocks [1, 30]. Incessant depleting rate and increasing cost of fossil fuel are factors of great concern to energy security [17, 18]. The energy security of the developing nations and non-oil-producing nations and the global economy in general is under siege [6]. Global energy demand is rising because of the exponential increase in populations and industrial growth. To realize the energy independence bearing in mind the world wide apprehension, it is indispensable to find eccentric fuel sources [23, 26]. Recently, over 350 famous oil-producing crops were identified as potential spring for biodiesel production [24].

Simulation is the imitation of the operation of a real-world process or system over time [2]. Simulation is used when the real system cannot be engaged, unaccessible and unacceptable to engage, to save time and cost [32]. The GT-Suite simulation software posses simulation modeling libraries-tools which is used to analyze engine combustion and acoustics, vehicle power trains, cooling systems, lubrication systems, fuel injection systems, valve trains and crankshafts [5]. Kyrtatos et al. [20] carried out a simulation and validation of combustion and pollutant at varying inlet conditions for common rail diesel engines. The research portrayed the ability of the model to predict the combustion and emissions of diesel fuel through varying the injection conditions and inlet charge in the diesel engines.

Hosseinzadeh et al. [7] conducted an numerical and experimental study of the effects of air filter holes masking on the altitude at heavy duty diesel engine (HDDE). Sinnamon [31] worked on co-simulation analyses of transient response and control for an engine with variable valve trains. Mohiuddin et al. [25] studied optimal design of automobile exhaust system using GT-Power. Exhaust (Manifold) system was designed using GT-Power software and its performance was compared with an existing system. It was established that, the newly designed exhaust manifold showed lower back pressure which ultimately resulted in better performance of the engine.

Liu et al. [21] established that the B100 biodiesel fuel reduced NO_x emissions by 10%. Despite the resourceful findings by the scholars on engine simulations, research on the simulation of the performance of biodiesel from various feedstocks in CI engines has left a vacuum in this broad area. Therefore, this article has investigated the performance of biodiesel from various feedstocks in a 4-cylinder diesel engine modeled in GT Suit. There are only a few researches that consider the simulation of binary biodiesel blends only, but less attention was given to biodiesel from pure feedstock. The objective of this research is to develop compression ignition engine model using GT-Power software, validate the model and simulate the engine performances of the biodiesel from various feedstocks under full loading condition at variable speeds.

2 Methodology

2.1 Materials, Equipment and Methods

2.1.1 Raw Materials

The raw materials used for the purpose of this research are as follows:

I. Diesel fuel sourced from Nigerian National Petroleum Corporation (NNPC) Mega Station, Katampe, Abuja
II. Biodiesel samples comprising 5 L each of cotton seed oil, jatropha seeds oil and neem seeds oil.

2.1.2 Equipment

i. A PC system installed with GT-Suite package version 7.4,
ii. The engine test bed used is a P8616 1600 CC 4-cylinder water cooled compression ignition engine manufactured by Cussons Technology, Manchester, and Year 2011 model. The Engine specifications are presented in Table 1.
iii. IMR-1400 gas analyzer, manufactured by Environmental Equipment Incorporation, 2014 Model, Central Avenue, Florida, USA.

2.1.3 Methodology

Modeling and Simulation of Engine Performance

See Fig. 1.

The one-dimensional numerical analyses of GT-Power software were used to simulate the diesel engine. Using the modeling tool of GT-Power, engine cycle simulation for biodiesel/petro diesel fuels and the engine specifications for the test rig used in Section "Secondary Biodiesel Data Sourcing and Analyses", a compression ignition engine was modeled (Built) into the GT-Power software to reflect the used test rig.

Table 1 Specifications of the engine fleet, manufactured by Cussons Technology, Manchester, UK, 2010

Engine type	4-cylinder in-line DOHC duel mass flywheel
Number of valves	16
Block	Aluminm
Head	Aluminm
Horse power	81 kW (11 OPS) at 4000 rpm
Torque	240–260 Nm at 1750 rpm with transient over boost
Displacement	1560 CC
Bore	75 mm
Stroke	88.3 mm
Compression ratio	18:1
Alignment	Transverse
Weight	107 kg
Dimensions $H \times L \times W$	$600 \times 500 \times 580$ mm

Source Cusson's manual (2010)

Equally important liquids information on density, enthalpy and transport properties is incorporated so that the properties of the liquid will be known if the fluid evaporates (Kofarbai and Zheng 2012). The modeled engine was simulated using the GT-Power software to study engine performance characteristics. These studies were made when the simulated engines were operating on petro diesel alone when the engines were running on biodiesels alone, as well as when they were running on various biodiesel blends with petro diesel. Plate 1 portrayed the GT-Power interface. Additionally, the simulation was conducted under full loading conditions and at an engine speed of 2500 rpm. The engines parameters like brake mean effective pressure; brake specific fuel consumption and brake thermal efficiency were recorded for each engine simulation [29]. After completing the simulation model, the results were recorded and tabulated for comparisons with those obtained from the engine fleet.

Secondary Biodiesel Data Sourcing and Analyses

The secondary data of the physicochemical properties of biodiesel made up of different feedstock such as yellow grease, palm, rapeseed, soy, canola, sunflower, tallow, corn, algae, coconut, camelina, safflower, waste cooking oil, castor, and others were obtained through and extensive literature review. Data from literature sources were obtained for most of the common biodiesel feedstock. For each physicochemical property per feedstock, the mean of the data, the mean deviation and the standard deviation was computed. The results were presented and discussed accordingly in comparison with ASTM and EN Standards. The statistical results were kept for further analyses [8].

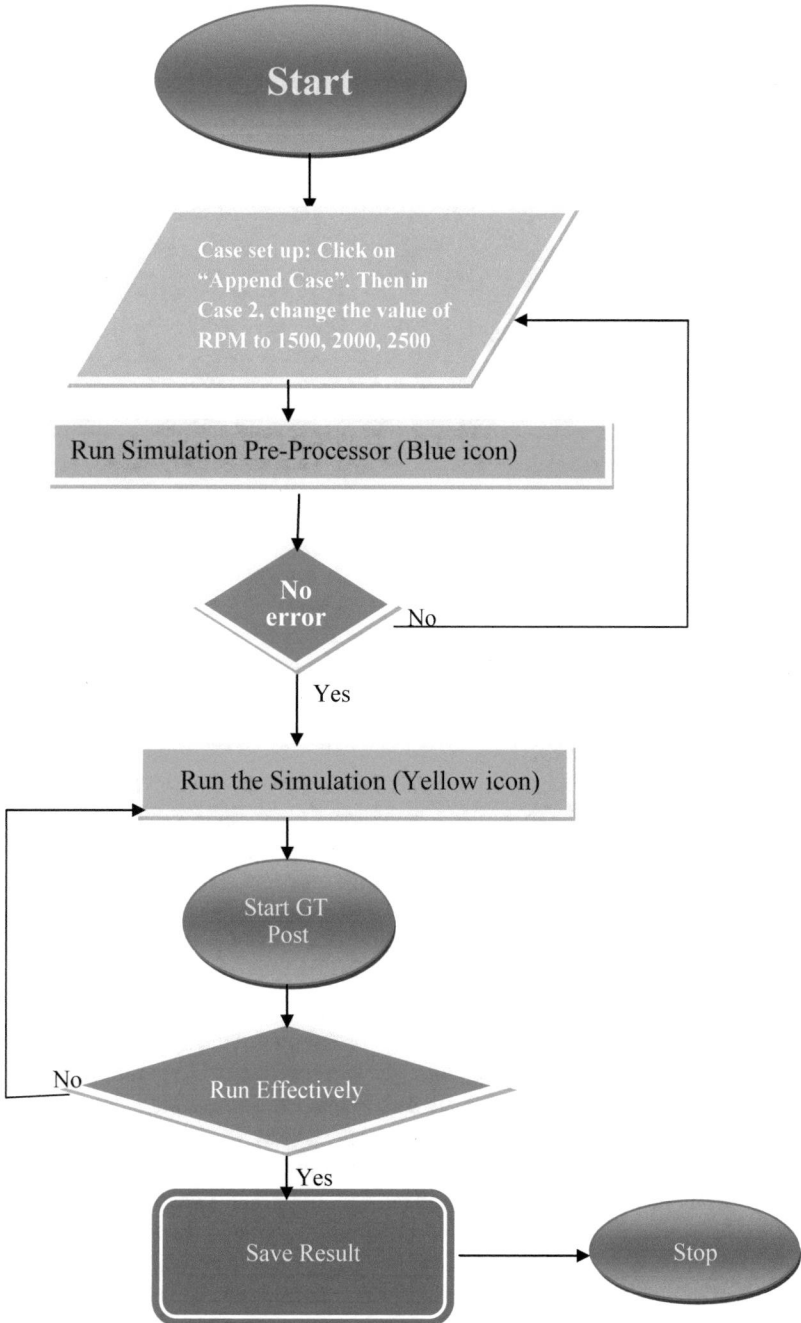

Fig. 1 Simple flowchart for the execution of simulation program

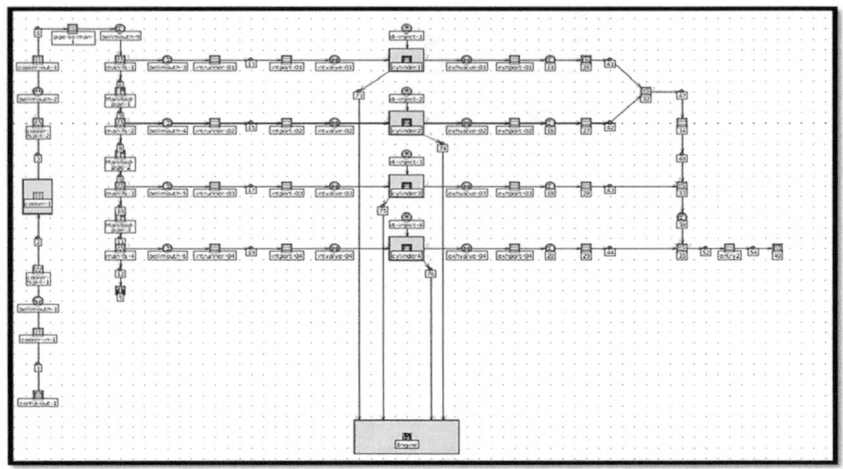

Plate 1 Four cylinder CI engine model from GT Suit

3   Results and Discussions

3.1   Results of Physicochemical Properties from Secondary Data Analyses

The means and deviation values for physicochemical properties of different biodiesel feedstock obtained from the literature are presented herewith in Tables 2 and 3. From the secondary data obtained, the mean kinematic viscosity values for all the biodiesel from various feedstocks under review except the castor oil biodiesel conform to the ASTM standards. Viscosity influences the ease of starting an engine, the spray quality, the particle size, and the quality of air-fuel mixing and the penetration of the injected jet.

The mean flash points of most biodiesel from various feedstock fall within ASTM limits. The flash points of soy, safflower and corn biodiesel are slightly higher, while that of coconut, algae and tallow are slightly lower than the ASTM min. The implication here is that; all the fuel samples are safe to handle. There storage is much safer; this is consistent with the works of Dandaje and Kaisan [3], and Kaisan et al. [10, 15, 16]. The mean heating values of all biodiesel except that of the castor confirm with EN minimum standards. All the mean specific gravity values except that the algae conform to ASTM minimum and maximum standards, most of the biodiesel values of pour point fall within ASTM limits except the palm and tallow biodiesel, with high positive values of 12.90 and 10.86 °C respectively. This is consistent with work of, Enwerenmadu et al. [4], and Kaisan et al. [18]. The specific gravity of a fuel has effects on the gravity of atomization and combustion and slightly affects fuel droplet size and air fuel mixing in an IC engine. All the average cetane values

Table 2　Mean physicochemical properties of biodiesel feedstocks [8]

Biodiesel Property	Yellow grease Mean	Dev	Palm Mean	Dev	Rapeseed Mean	Dev	Soy Mean	Dev	Canola Mean	Dev	Sun flower Mean	Dev	Jatropha Mean	Dev	Tallow Mean	Dev	Corn Mean	Dev	Algae Mean	Dev
Viscosity (mm²/S)	4.71	0.28	4.91	0.56	4.45	0.24	4.33	0.23	4.40	0.27	4.60	0.39	4.94	1.00	4.67	0.31	3.97	0.51	4.32	0.72
Density (g/cm³)			0.87	0.008	0.89	0.014	0.89	0.014	0.88	0.00	0.854	0.04	0.88	0.01			0.90	0.022	0.95	0.029
Flash point (°C)	139.30	29.98	165.33	16.39	166.63	29.84	183.14	33.22	140.86	24.16	169.33	24.49	158.59	26.91	123.57	31.51	192.00	34.00	126.63	36.13
Acid value			0.11	0.10			0.10	0.06	0.17	0.16	0.18	0.09	0.44	0.06					0.45	0.04
HV (MJ/kg)	39.01	0.83	37.90	1.91	40.16	1.96	38.74	1.60	39.56	1.51	40.07	1.90	42.77	3.08	38.90	1.00	41.47	2.36	39.36	2.79
Cetane number	54.42	3.05	62.91	2.32	51.36	3.54	50.18	4.37	56.05	3.63	50.86	3.93	55.96	3.33	60.09	3.45	53.22	6.25	51.00	1.33
Pour point (°C)	−0.50	4.25	12.90	1.34	−11.50	4.40	−4.00	2.62	−8.14	2.43	−2.80	2.24	−0.50	4.00	10.86	2.65	−2.50	1.50	−10.50	0.00
Cloud point (°C)	5.71	5.47	14.69	1.30	−3.50	1.13	0.38	1.49	−1.88	1.88	1.44	2.05	3.20	3.84	13.75	2.19	−3.00	0.00	−11.50	0.00
CFPP (°C)	1.54	4.19	10.43	2.98	−13.44	4.49	−3.67	2.37	−7.75	1.75	−2.80	1.76	0.00	0.00	12.50	1.50	−7.50	4.50	−10.00	0.00

(continued)

Table 2 (continued)

Biodiesel Property	Yellow grease		Palm		Rapeseed		Soy		Canola		Sun flower		Jatropha		Tallow		Corn		Algae	
	Mean	Dev	Mean	Dev	Mean	Dev	Mean	Dev	Mean	Dev	Mean	Dev	Mean	Dev	Mean	Dev	Mean	Dev	Mean	Dev
Copper strip corr.																				
Sulfur cont (ppm)	3.91	3.34	3.38	2.91	2.73	1.11	2.10	1.72	2.00	0.50	2.33	1.55	0.60	0.48	15.20	14.24	4.00	1.00		
Specific gravity	0.88	0.01	0.88	0.00	0.88	0.01	0.88	0.00	0.88	0.00	0.88	0.01	0.88	0.00	0.88	0.00	0.88	0.00	0.91	0.00
Sulfated ash (%)									0.01	0.00			0.06	0.00						
Carbon residue (%)					0.05	0.03					0.08	0.07	0.20	0.00						
Iodine value	87.78	13.18	51.67	6.10	112.00	1.60	123	4.91	110.17	1.74	104.04	28.43	41.80	3.40	54.10	15.27	101.00	0.00	64.50	7.50
No. of references	20		20		20		20		20		20		20		14		8		8	

Table 3 Mean physicochemical properties of biodiesel feedstocks (continuation) [8]

Biodiesel	Coconut		Camelina		Safflower		WCO		Castor		Cotton		Others	
Property	Mean	Dev.	Mean	Dev.	Mean	Dev.	Mean	Dev.	Mean	Dev.	Mean	Dev.	Mean	Dev.
Viscosity (mm²/S)	2.95	0.33	4.26	0.80	4.14	0.10	4.83	0.38	11.44	1.16	4.48	0.03	5.02	1.04
Density (g/cm³)							0.88	0.005	0.89	0.02	0.88	0.003		
Flash point (°C)	113.00	4.50	136.0	0.00	174.3	4.89	164.0	0.00	151.5	15.75	169.0	1.67	169.8	28.17
Acid value			45.10	0.00			0.16	0.09	0.35	0.13	0.17	0.01		
HV (MJ/kg)	37.13	1.29	45.30	0.30	42.20	2.00			30.40	0.60			39.91	1.38
Cetane number	59.33	7.11	50.43	1.19	51.05	1.25	61.00	0.00					54.50	4.11
Pour point (°C)	−8.50	3.50	−7.20	2.56	−7.33	0.89	−13.00	3.00					1.00	5.00
Cloud point (°C)	−2.67	1.78	2.67	0.44	−3.50	1.50	−1.00	0.00	−13.00	0.67			8.00	0.00
CFPP (°C)	−4.50	0.50	−3.00	1.00	−6.00	0.00	−5.30	0.00						
Copper strip corr.									1.00	0.00				
Sulfur cont (ppm)	2.67	0.44	2.00	1.00			2.00	0.00					9.00	0.00
Specific gravity	0.88	0.01	0.88	0.01	0.88	0.01							0.88	0.00
Sulfated ash (%)							0.01	0.00						
Carbon residue (%)														
Iodine value	30.00	0.00	152.8	1.75	141.0	0.00	105.0	0.00					98.78	21.48
Number of references	9		7		4		5		4		6		15	

under consideration conformed to the ASTM standards. Generally, shorter ignition delay results in higher cetane number and vice versa. This implies that all the blends herein examined would have the shortest possible ignition delay in a CI engine.

3.2 Exhaust Emissions Results

The results of the engine emissions at the engine speed of 1000, 1500, 2000 and 2500 rpm are presented in Tables 4, 5, 6 and 7.

From Tables 4, 5, 6 and 7, the CO and NO_x emissions of the three biodiesels and diesel were portrayed at engine speeds of 1000, 1500, 2000 and 2500 rpm respectively.

From the tables, the CO emission of cotton biodiesel maintained a constant value of 0.04 g/m^3 at all engine speeds. Jatropha biodiesel showed a common value of 0.05 CO emission in g/m^3 between 1000 and 1500 rpm, at 2000 rpm, the value increased to 0.06 g/m^3 and at the highest speed of 2500 rpm showed a further increase to 0.08 g/m^3. This is contrary to the carbon monoxide emissions of neem which increased steadily from 0.47, 0.55, 0.559 and 0.8 g/m^3 at engine speeds of 1000, 1500, 2000 and

Table 4 Exhaust emissions for biodiesels from cotton, jatropha and neem at 1000 rpm

Flue gases	Cotton B100	Jatropha (B100)	Neem (B100)	Diesel (B000)
CO (g/m^3)	0.04	0.05	0.47	0.81
NO_x (g/m^3)	0.081	0.24	0.25	0.267

Table 5 Exhaust emissions for biodiesels from cotton, jatropha and neem at 1500 rpm

Flue Gases	Cotton B100	Jatropha (B100)	Neem (B100)	Diesel (B000)
CO (g/m^3)	0.04	0.05	0.552	0.863
NO_x (g/m^3)	0.084	0.24	0.26	0.278

Table 6 Exhaust emissions for biodiesels from cotton, jatropha and neem at 2000 rpm

Flue gases	Cotton B100	Jatropha (B100)	Neem (B100)	Diesel (B000)
CO (g/m^3)	0.04	0.06	0.559	1.25
NO_x (g/m^3)	0.12	0.24	0.3	0.358

Table 7 Exhaust emissions for biodiesels from cotton, jatropha and neem at 2500 rpm

Flue gases	Cotton B100	Jatropha (B100)	Neem (B100)	Diesel (B000)
CO (g/m^3)	0.04	0.08	0.8	0.81
NO_x (g/m^3)	0.15	0.24	0.32	0.36

2500 rpm respectively. The diesel was leading in each case in terms of CO emission. Largely, the partial combustion of fuel and the Carbon monoxide (CO) emissions are due to insufficient molecular oxygen content in a fuel [24]. In general, factors such as air-fuel ratio; speed of the engine, injection time and pressure and the type of fuel used influence the CO discharge. The zenith values of Carbon Monoxide emission were given by the petrol diesel. Neem biodiesel is not too good in terms of CO emissions. The NO_x emissions from Tables 4, 5, 6 and 7 showed that cotton has the least emission at all engine speeds. The maximum value of Nitrogen Oxides (NO_x) emission was exhibited diesel fuel. Astonishingly, B100C, B100N and B100J gave relatively the low NO_x emission tendencies. The value of NO_x emitted by neem is higher than cotton and jatropha. This is because neem has more Nitrogen in its seeds than the duo, this accounts for the use of neem as a substratum for organic fertilizer production. The results are in conformity with. Although, they are inconsistent with Ong et al. [28, 27].

3.3 Modeling and Simulation Using GT-Power Software

3.3.1 Model Validation

The simulated results of the engine performance were compared with the result of the engine test obtained, and presented in Figs. 2, 3 and 4 for the specific fuel consumption, brake mean effective pressure and brake thermal efficiency respectively. The pictures of the result window sample displayed on the GT-Post (From the GT-Suite software) are shown in Plate 2.

The experimented and simulated specific fuel consumption of jatropha. Cotton and neem biodiesel during respective speeds of 1000, 1500, 2000 and 2500 rpm are shown in Fig. 2. It could be observed that, there is no significant difference between the experimental values and the simulated values. The BSFC at 1000 rpm for cotton biodiesel has a difference of 1.37% been the lowest difference between the experimented and simulated values, while the BSFC of neem biodiesel at 2000 rpm has a difference of 13.75% been the highest difference between the experimented and simulated values. The remaining values are in between the former two and are less than 10% each, which makes them acceptable in accordance with the findings. The BSFC for the experimental and simulated data show relative correspondence in each case and at all speeds. The simulated results increase with corresponding increase in experimental results and reduce with corresponding decrease in experimented results in all cases of specific fuel consumption. This makes the GT-Post findings to be real since they are very consistent with the experimented data.

Furthermore, Fig. 3 depicts the experimented and simulated values of brake mean effective pressure of jatropha, cotton and neem biodiesel at respective speeds of 1000, 1500, 2000 and 2500 rpm. It can be perceived that, there is no significant difference between the experimental values and the simulated values. The bmep at 2500 rpm for cotton biodiesel has a difference of 1.44% been the lowest difference

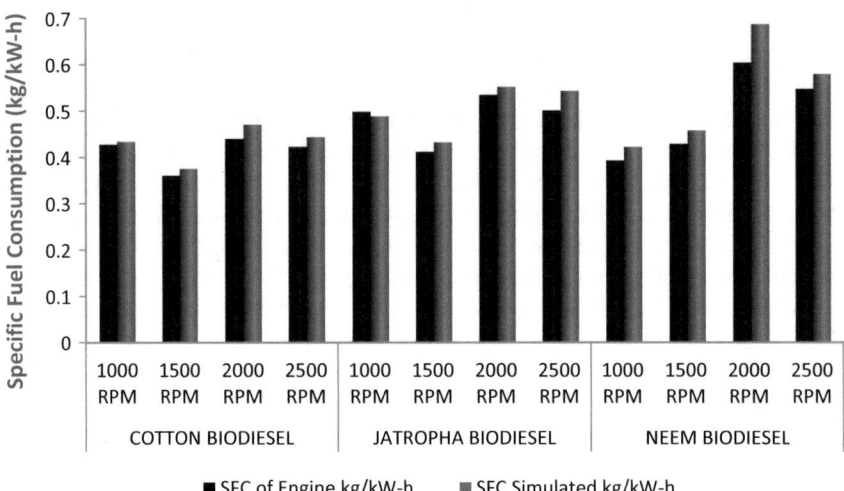

Fig. 2 Comparison of specific fuel consumption of experimental and simulated results

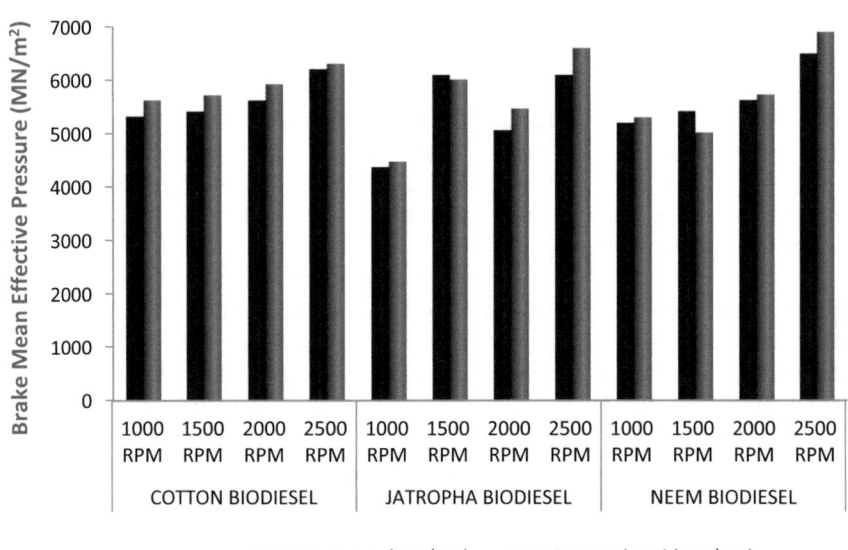

Fig. 3 Comparison of brake mean effective pressure of experimented and simulated results

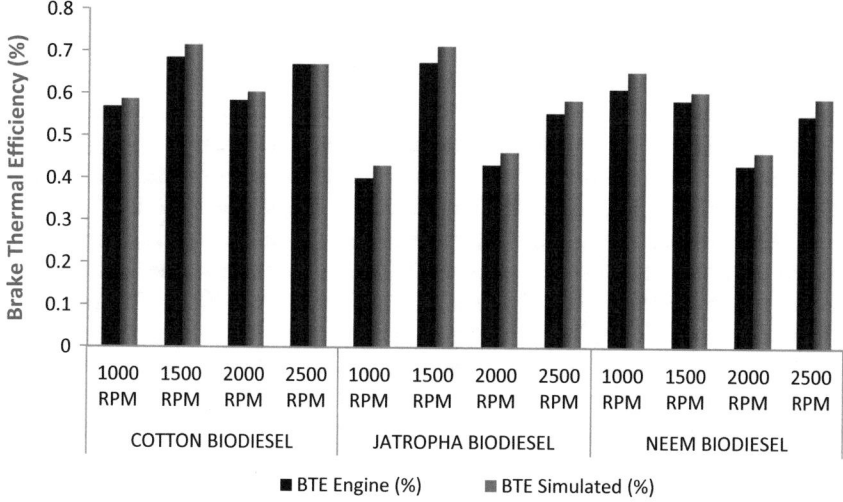

Fig. 4 Comparison between brake thermal efficiencies of experimental and simulated results

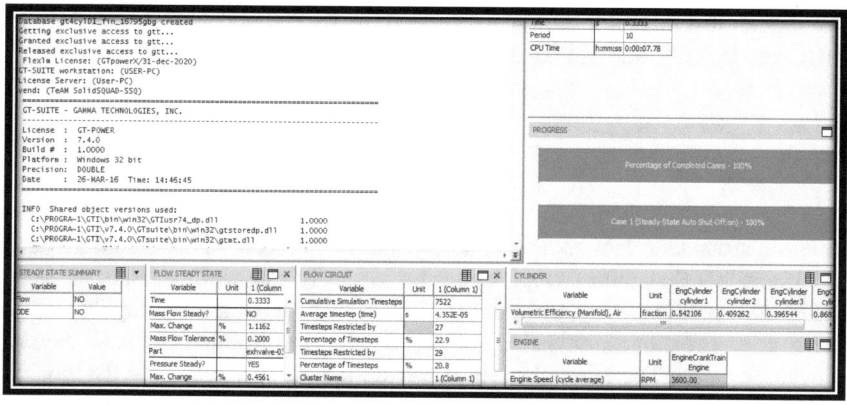

Plate 2 Sample simulation result in GT-Post

between the experimented and simulated values, while the bsfc of jatropha biodiesel at 1500 rpm has a difference of 8.20% been the highest difference between the experimented and simulated values. The remaining values are in between the former two and four of them are just slightly above 1% difference each, this makes them acceptable in accordance with the findings. The bmep for the experimented and simulated data show relative correspondence in each case and at all speeds. The simulated results increase with corresponding increase in experimented results and reduce with corresponding decrease in experimented results in all cases of specific fuel consumption. This makes the GT-Post findings to be real since they are very consistent with the experimental data.

Similarly, Fig. 4 illustrates the experimented and simulated values of brake thermal efficiency of cotton, jatropha and neem biodiesel at respective speeds of 1000, 1500, 2000 and 2500 rpm. Just like the two earlier cases, there are no any significant differences between the experimented and simulated data in brake thermal efficiencies of biodiesels from cotton, jatropha and neem at all speeds. At the speed of 2500 rpm, for cotton biodiesel, the brake thermal efficiencies are exactly the same values for both experimented and simulated results. This can be explained by the fact that, engine performance models are often dependent on the experimental setup in which they were determined, thus giving slightly different results depending on the context where they are used (Abbe et al. 2013; Kolhe et al. 2015). This makes the GT-Suite very reliable software in engine performance simulations. Therefore, the software can be used to determine the performance of CI engines with any biodiesel fuel at any speed. Statistical analyses confirmed that, the mean value of the experimental brake specific fuel consumption was 0.465 kg/kW-h and the mean value for the simulated brake specific fuel consumption was 0.491 kg/kW-h. The standard deviation in brake specific fuel consumption between the experimented and simulated values was 0.078. A similar trend was shown in the values of brake thermal efficiencies. The experimental brake thermal efficiency values gave an average 0.563 while the simulated values gave a mean of 0.589. The standard deviation was 0.094. This implies that there is a strong tie between the experimental and simulated values and hence, the GT-Suite can be used to replace a typical engine rig.

4 Conclusion

From the secondary data obtained all the feedstocks reviewed have conformed to ASTM and EN Standards except: castor for viscosity and heating value; algae for density and specific gravity; and palm and tallow for pour point. This implies that the quality of combustion, atomization and fuel droplets can be improved in CI engines using these blends Therefore; the fuel properties indicate that they have the possible to replenish the energy burden in an eco-friendly method and its likelihood to be a viable fuel source for CI engines. Biodiesel displays positive emissions level and emits fewer pollutants than that of diesel. Pure biodiesel extracted from jatropha, cotton and neem give the lower CO and NO_x emissions compared with the petro diesel. Cotton biodiesel has the same value of CO emissions which is 0.04 g/m^3 throughout the experiment at all engine speed as well as during simulation. The least values of CO emission were 0.03 g/m^3 both for Sunflower and Safflower biodiesel which also had the lease values of NO_x emissions as 0.11 and 0.14 g/m^3.

References

1. Ashok, B., Nanthagopal, K., Kavalipurapu, R. T., Saravanan, B. D., Arumuga P., & Kaisan, M. U. (2018). Experimental study of methyl tert-butyl ether as an oxygenated additive in diesel and *Calophyllum inophyllum* methyl ester blended fuel in CI engine. *Environmental Science and Pollution Research.* https://doi.org/10.1007/s11356-018-3318-y. Published by Springer.
2. Banks, J., Carson, J., Nelson, B., & Nicol D. (2001). *Discrete-event system simulation* (Vol. 1, 3, pp. 1–3). Prentice Hall.
3. Dandajeh, H. A., & Kaisan, M. U. (2014). Engine performance characteristics of a gardener compression ignition engine using rapeseed methyl ester. In *Nigeria Engineering Conference*, Ahmadu Bello University, Zaria, Nigeria.
4. Enwerenmadu, C. C., Omodolu, C., Mustapha, T., & Rutto, L. (2014). Effects of feedstock-related properties on engine performance of biodiesel from canola and sunflower oils of South African Origin. *Journal of Industrial and Mechanical engineering*, 15–16.
5. GTI Soft. (2014).*GT Power engine simulation software*. Retrieved on 14th August, 2014 from http://www.gtisoft.com/applications/a_Engine_Performance.php.
6. Giakoumis, E. G. (2012). A statistical investigation of biodiesel effect on regulated exhaust emission during transient cycles. *Applied Energy, 98,* 273–291.
7. Hosseinzadeh, S., Bandpy, M. G., Rad, G. J., & Keshavarz, M. (2012). Experimental and numerical study of impact of air filter holes making on altitude at heavy-duty diesel engine. *Modern Mechanical Engineering, 2* , 157–166. http://www.eia.gov/forecasts/steo/pdf/steo_full.pdf.
8. Kaisan, M. U. (2017). *Modelling and simulation of biodiesel blends into compression ignition engine*. Unpublished Ph.D. Research, Ahmadu Bello University, Zaria, Nigeria.
9. Kaisan, M. U., Abubakar, S., Umaru, S., Dhinesh, B., Mohamed Shameer, P., Mohamed Nishath, P. (2018). Comparative analyses of biodiesel produced from jatropha and neem seeds oil using gas chromatographic and mass spectroscopic technique. *Biofuels,* 1–12. https://doi.org/10.1080/17597269.2018.1537206.
10. Kaisan, M. U., Anafi, F. O., Nuszkwoski, J., Kulla, D. M., & Umaru, S. (2016). *Review of physico-chemical properties of biodiesel feedstock and their suitability for Nigerian economy*. Oriental Hotel, Lagos, Nigeria: Nigerian Energy Forum.
11. Kaisan, M. U., Anafi, F. O., Nuszkowski, J., Kulla, D. M., & Umaru S. (2017). Calorific value, flash point and cetane number of biodiesel from cotton, jatropha and neem oil binary and multi-blends with diesel. *Biofuels, 8*(4), 1–7. Published by Taylor and Francis.
12. Kaisan, M. U., Anafi, F. O., Nuszkwoski, J., Kulla, D. M., & Umaru S. (2017). Exhaust emissions of biodiesel binary and multi-blends from cotton, jatropha and neem oil from stationary multi cylinder CI engine. *Transportation Research Part D: Transport and Environment, 53*(2017), 403–414. Published by Elsevier.
13. Kaisan, M. U., Anafi, F. O., Nuszkwoski, J., Kulla, D. M., & Umaru S. (2017). Fuel properties of biodiesel from cotton, jatropha binary and multi-blends with diesel. *Nigerian Journal of material Science and Engineering, 7*(1): 55–61. Published by Material Science and Technology Society of Nigeria.
14. Kaisan, M. U., Anafi, F. O., Nuszkwoski, J., Kulla, D. M., & Umaru S. (2017). Cold flow properties and viscosity of biodiesel from cotton, jatropha and neem binary and multi-blends with diesel. *African Journal of Renewable and Alternative Energy, 1*(1), 80–89. Published by Renewable and Alternative Energy Society of Nigeria.
15. Kaisan, M. U., Anafi, F. O., Nuszkwoski, J., Kulla, D. M., & Umaru S. (2016). Fuel properties of biodiesel from cotton, jatropha binary and multi-blends with diesel. Material Science and technology Society of Nigeria. NARICT, July, 2016.
16. Kaisan, M. U., Anafi, F. O., Nuszkwoski, J., Kulla, D. M., & Umaru, S. (2016). Performance evaluation of binary and multi-blends of biodiesel from cotton, jatropha and neem oils in multi cylinder CI engine. In: *International Conference on Alternative Fuels*, University of Kayseri, Turkey.

17. Kaisan, M. U., Nafiu, T., & Habib, Y. B. (2010). Carbon capture storage and processing as a means of enhancing renewable energy sources in Nigeria. *Renewable and Alternative Energy for Sustainable National Development, 2*(1), 219–225.

18. Kaisan, M. U., Naifu, T., & Habib, Y. B. (2010). Towards new policies in minimizing green house gas emission in Nigeria. *Renewable and Alternative Energy for Sustainable National Development, 2*(1), 306–316.

19. Kaisan, M. U., Pam, G. Y., Kulla, D. M., & Kehinde, A. J. (2015). Effects of Oil Extraction Method on Biodiesel Production from Wild Grape Seeds: A Case Study of Soxhlet Extraction Method and Mechanical Press Engine Driven Expeller Method. *Journal of Alternate Energy and Technologies, 6*(1), 35–41.

20. Kyrtatos, P., Obrecht, P., Hoyer, K., & Boulouchos, K. (2010). Predictive simulation and experimental validation of phenomenological combustion and pollutant models for medium-speed common rail diesel engines at varying inlet conditions. *International Council on Combustion Engines, 143,* 1–13.

21. Liu, H., Strank, S., Werst, M., Hebner, R., & Osara, J. (2010). Combustion emissions modelling and testing of conventional diesel fuel. In *ASME 2010 4th International Conference on Energy Sustainability. Phoenix, AZ, USA.* ES2010-90037, 1–10.

22. Lomonaco, D., Maia, F. J. N., Clemente, C. S., Mota, J. P. F., Costa Junior, A. E., & Mazzetto, S. E. (2012). Thermal studies of new biodiesel antioxidants synthesized from a naturally occurring phenolic lipid. *Fuel, 97,* 552–559.

23. Mofijur, M., Atabani, A. E., Masjuki, H. H., Kalam, M. A., & Masum, B. M. (2013). A study on the effects of promising edible and non-edible biodiesel feedstocks on engine performance and emissions production. A comparative evaluation. *Renewable and Sustainable Energy Reviews, 23,* 391–404.

24. Mofijur, M., Masjuki, H. H., Kalam, M. A., Atabani, A. E., Rizwanal Fattah, I. M., & Mobarak, H. M. (2014). Comparative evaluation of performance and emission characteristics of *Moringa oleifera* and palm oil based biodiesel in a diesel engine. *Journal of Industrial Crops and Product., 53,* 78–84.

25. Mohiuddin, A. K., Ataur-Rahman, M., & Dzaidin, M. (2007). Optimization design of automobile exhaust system using GT-Power. *International Journal of Mechanical and Material Engineering, 2*(1), 40–47.

26. Nafiu, T., Magaji, U. I., Zuru, A. A., Kaisan, M. U., & Habib, Y. B. (2011). Production of biodiesel from wild grape seeds. *Technical Transaction Journal of Nigerian Society of Engineers, 46*(1), 1–10. Published by Nigeria Society of Engineers.

27. Ong, H. C., Masjuki, H. H., Mahlia, T. M. I., Silitonga, A. S., Chong, W. T., & Leong, K. Y. (2014). Optimization of biodiesel production and engine performance from high free fatty acid *Calophyllum inophyllum* oil in CI diesel engine. *Energy Conversion and Management, 81,* 30–40.

28. Ong, H. C., Silitonga, A. S., Masjuki, H. H., Mahlia, T. M. I., Chong, W. T., & Boosroh, M. H. (2013). Production and comparative fuel properties of biodiesel from non-edible oils: *Jatropha curcas, Sterculia foetida* and *Ceiba pentadra. Energy Conversion and Management, 73,* 243–255.

29. Rahim, R., Mamat, R., Taib, M. Y., & Abdullah, A. A. (2012). Influence of fuel temperature on a diesel engine performance operating with biodiesel blend. *International Journal of Advance Sciences and Technology, 43,* 115–126.

30. Sani, S., Kaisan M. U., Kulla, D. M., Ashok, B., & Nanthagopal, K. (2018). Determination of physico chemical properties of biodiesel from *Citrullus lanatus* seeds oil and diesel blends. *Industrial Crops and Products, 122,* 702–708. https://doi.org/10.1016/j.indcrop.2018.06.002. Published by Elsevier.

31. Sinnamon, J. F. (2007). Co-simulation analysis of transient responses and control for engines with variable valve trains. *SAE International, 1223,* 1–15.

32. Sokolowski, J. A., & Banks, C. M. (2009). *Principles of modeling and simulation hoboken* (pp. 6–7). NJ: Wiley.

Modelling and Simulation of Biodiesel from Various Feedstocks into Compression Ignition Engine

Muhammad Usman Kaisan, Shitu Abubakar, Fatai Olukayode Anafi, Samaila Umaru, P. Mohamed Shameer, Umar Ali Umar, Sunny Narayan, P. Mohamed Nishath and J. Senophiyah Mary

Abstract This article has described how a 4-cylinder compression ignition (C.I.) engine model can be developed using the GT-Suite software interface with known specifications of an existing engine fleet. Fuel properties of various grades of biodiesel were computed and their mean values were fed into the simulated model. The simulation of engine in GT-Power established that biodiesel fuels produced lower amounts of carbon monoxide, nitrogen oxide and Sulphur dioxide emissions as compared to diesel. The Sulphur dioxide (SO_2) emissions have no significance for all the biofuels. The simulation of the secondary fuel data revealed that highest brake specific fuel

M. U. Kaisan (✉) · S. Abubakar · F. O. Anafi · S. Umaru · U. A. Umar
Department of Mechanical Engineering, Faculty of Engineering, Ahmadu Bello University, Zaria, Nigeria
e-mail: mukaisan@abu.edu.ng

S. Abubakar
e-mail: abubakarshitu88@gmail.com

F. O. Anafi
e-mail: fataianafi@yahoo.com

S. Umaru
e-mail: bnumar@yahoo.com

U. A. Umar
e-mail: umaraliumar@yahoo.co.uk

P. Mohamed Shameer
Department of Mechanical Engineering, V. V. College of Engineering, Tirunelveli 627657, Tamil Nadu, India
e-mail: pmohamedshameer@gmail.com

S. Narayan
Department of Mechanical Engineering, Faculty of Engineering, Qassim University, Buraydah, Saudi Arabia
e-mail: rarekv@gmail.com

P. Mohamed Nishath · J. Senophiyah Mary
Government College of Technology, Coimbatore 641013, India
e-mail: pmdnishath@gmail.com

J. Senophiyah Mary
e-mail: senophiyah.mary8@gmail.com

© Springer Nature Singapore Pte Ltd. 2020
S. K. Ghosh (ed.), *Energy Recovery Processes from Wastes*,
https://doi.org/10.1007/978-981-32-9228-4_9

consumption (BSFC) was shown by castor oil followed by coconut and tallow oil. The highest brake mean effective pressure was achieved during the simulation of performance of CI engine using sun flower oil, tallow oil, waste cooking oil and soy based biodiesels correspondingly. The highest brake thermal efficiency (BTE) values were depicted by soy based biodiesel, sun flower oil, palm oil and safflower oil in descending order.

Keywords Biodiesel · GT-power · Modelling · Simulation · CI engine

1   Introduction

Biodiesel is a fatty acid of alkyl esters produced through catalysed transesterification reaction of alkanols with vegetable oils and animal fats [1]. Biodiesel can be defined as the mono alkyl ester of long-chain fatty acids produced from sustainable biological feedstocks [2]. Previous research works have been aimed at finding lasting solutions for global fuel crises and environmental impacts associated with consumption of various fossil fuels. Some of the related research works are presented in this section. Kolade et al. [3] has presented one-dimensional and three-dimensional injection processing diesel engine. It has been established that, the injection process not only effects the combustion process, but also properties like efficiency of the fuel, combustion noise, emissions, and acceleration of engine. Bhave et al. [4] worked on modelling of a dual fuelled, multi-cylinder, homogenous charged compression ignition engine using an engine cycle simulator. They examined effects of octane number on engine performance, combustion characteristics and engine emissions for case of a multi-cylinder truck engine. A homogenous charged compression ignition engine was fuelled using straight chain heptane, branched chained heptane and branched chained octane based fuels. The authors observed a commendable bond amongst the forecasted values and experimented values for auto-ignition, in cylinder pressure developed, and emissions. Rahim et al. [5] examined impacts of fuel injection temperatures on performance of a diesel engine working on a 5% biodiesel/diesel blend. They measured typical physical properties of the fuel used before it could be installed into GT-Power data base. The standard fuel database of GT-Power software did not contain any biodiesel property. Rahim et al. [6] have done a relative study of a one-dimensional simulation of biodiesel blended with diesel. The pure vegetable oil, 20% blended biodiesel and a 5% blended biodiesel were used as fuel respectively. Fuel properties of GT-Power software simulation software were used to get comparative results of engine boundaries. There was no substantial variance in the performance of the engine when 5% biodiesel and 100% diesel fuels were used. Kakee and Pishgoogie [7] determined the optimum valve operational timings for various spark ignition engines using constraints obtained from various simulation methods. A four-stroke spark ignition engine was modelled using GT-Power software and linked to the Simulink and MATLAB in order to regulate various input/output boundaries. The discussed model was further used for analysis of sensitivity and engine specific fuel consumption. An improvement in engine performance was recorded, which was established by smoother torque diagram in the VVT. Also, torque variations increased with engine speed at lower values of engine speeds.

The main objective of this work is to evaluate the performance of a 4-cylinder compression ignition engine operating under full loading condition at variable speeds, using various types of biodiesel stocks using modelling in GT Suit.

2   Literature Review

This part of work presents an overview of previous works done to study the performance of various types of engines using biofuel/biofuel blends as fuel. Ren et al. [8] studied physical and chemical properties of biodiesel based fuels on the combustion process and pollutants formation in direct injection (DI) engines using multidimensional computational fluid dynamics (CFD) simulation. While the variations of various properties influence the formation of soot and NO_x emissions. Gracianno et al. [9] analysed a model for compression ignition internal combustion engines driven by diesel, biodiesel, and/or biogas. Results have shown that using pure biodiesel as a fuel reduced power output by about 1.0%, and at the same time the consumption of fuel rose by about 12.0% when compared to diesel as a fuel. Furthermore, the use of natural gas in CI engines reduce engine power output and fuel consumption by 2.0%, and 13.0% respectively when compared to fossil diesel as a fuel. The presented model can be an effective tool in order to design, control, and optimization performance of compression ignition engines. Aldhaidhawi et al. [10] proposed an engine model that could predicate the performance and exhaust gas emissions of a single cylinder four-stroke direct injection engine fuelled with diesel and palm oil methyl ester of B7 (blends 7% palm oil methyl ester with 93% diesel by volume) and B10 biodiesel fuel. Results confirm that biodiesel obtained from palm oil methyl ester could represent a attractive alternative to using pure diesel fuel when engine is operated with variable compression ratios. A comprehensive computer code using "C" language was developed for analysing engine cylinder pressure, heat release, heat transfer and performance characteristics such as work done, brake power and brake thermal efficiency (BTE) of a compression ignition (C.I.) engine [11]. It was observed that by using the blend B20 as the fuel, the peak pressure and temperature increases and hence resultant stresses [11]. GT-Power software code was used for simulation of a biodiesel-fuelled direct injection compression ignition engine cycle to study its performance and emission characteristics [12]. Results from sensitivity analysis proved that timings of fuel injection and EGR percentage had the most significant effect on formation of nitric oxides [12]. Morse simulated the combustion reaction within an optical Sandia/Cummins N14 direct injection compression ignition engine [13]. The negative effects of biodiesel fuel impingement on the piston and wall were analysed. Boggavarapu and Ravikrishna [14] studied the effect of physicochemical properties on spray atomization for both biofuels and fossil fuels. Evaporating sprays revealed that the liquid length is longer for biodiesel. Nasim et al. [15] analysed performance of a compression ignition engine powered by neat jatropha oil. Thermal efficiency of engine decreased when it was powered by preheated neat jatropha oil at higher speeds. Harch et al. [16] studied the combustion model of a C.I. engine using AVL Fire software. The fuel used was 5% BLS biodiesel (B5) and 10% BLS biodiesel (B10) for different injection timings and compression ratios. Use of B10 biodiesel provided performance and efficiency of engine. Al-Dawody et al. [17] used Wiebe function to calculate the instantaneous heat release rate and results were validated through Diesel-RK workbench.

3 Simulations

3.1 GT-Power

This is a type of standard engine software model well adapted by all leading engine manufacturers, and producers. This software is widely used in power generating engines, ship construction, small engines as well as racing engines [18]. The simulation in GT-Suite comprises of a set of modelling libraries which are used for evaluating engine sentient, valve trains, power trains, ignition and audibility, engine cooling structures, crankshafts, fuel injection systems, and lubrication systems.

4 Methodology

The engine under test was a P8616 1600 Cc 4-cylinder water cooled compression ignition engine manufactured by Cussons Technology. The specifications of the test rig model are portrayed in Table 1.

4.1 Secondary Biodiesel Data Sourcing and Analyses

Data from literature sources were obtained for properties of the biodiesel feedstocks. For each physicochemical property per feedstock, the average values of the biofuel property, the standard deviations and the mean deviation were computed. The results were presented and discussed accordingly in comparison with ASTM and EN Standards [19].

Table 1 Stipulations of the engine model	Engine type	4-Cylinder in-line duel mass flywheel
	Horse power	81 kW at 4000 rpm
	Block	Aluminium
	Number of valves	16
	Displacement	1560 CC
	Torque	240–260 NM
	Compression ratio	18–1
	Alignment	At right angles
	Stroke	88.3 mm
	Bore	75 mm
	Dimensions H × L × W	600 × 500 × 580 mm
	Weight	107 kg

4.2 Modelling and Simulation of Engine Performance

The one-dimensional numerical analyses of GT-Power software have been used to simulate the diesel engine data. According to reference in GT-Power, exhaust gas properties like carbon, hydrogen, oxygen and nitrogen contents as well as calorific value, enthalpy, critical pressure, critical temperature and transport properties like thermal conductivity and dynamic viscosity were recorded. Equally important liquids information about density, enthalpy and transport properties are incorporated [20].

The modelled engine was simulated using petro-diesel, biodiesels, and various biodiesel blends with petro diesel as fuel. Plate 1 portrayed in Fig. 1 shows the GT-Power interface.

Additionally, the simulation was conducted under full loading conditions and operating engine at speed of 2500 rpm. The engines performance factors like brake thermal efficiency (BTE), brake specific fuel consumption (BSFC) and brake mean effective pressure (BMEP) were documented for each engine simulation [21]. After completing the simulation model, the results were recorded and tabulated for comparisons with those obtained from the engine fleet. After the software validation, the same engine fleet was simulated using different fuel properties of biodiesels obtained from the secondary data. The list of biodiesel fuels used include: yellow grease based sample, palm based sample, rapeseed based sample, soy based sample, canola based sample, sunflower oil sample, tallow sample, corn sample, algae sample, coconut sample, camelina sample, safflower sample, waste cooking oil sample, castor oil sample, and others (Peanut, karanja, rice bran, sorghum, lard, poultry fat, mustard, peanut, babarsu, fish oil, kusum, beheda and groundnut). Conclusively, the results obtained were presented, analysed, discussed and concluded accordingly. The GT-Power software was launched a 4-cylinder engine model was developed in the software interface. Four test cases were created for the engine speeds of 1000, 1500, 2000 and 2500 rpm respectively. The simulation was repeated for the biodiesel feed-

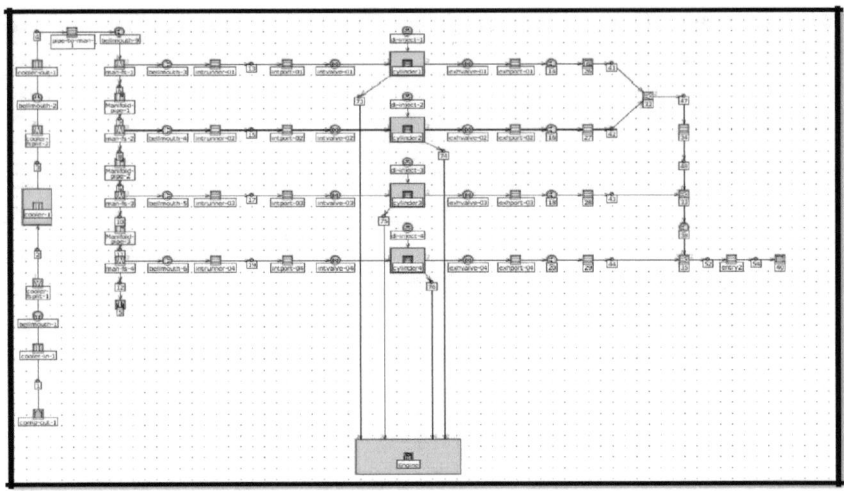

Fig. 1 Plate 1—engine model from GT-Power software

stock sourced from the secondary data using the fuel properties of these feedstocks. The results of the engine parameters with CO and NO_x remained logged and kept for further analyses. Figure 2 shows the flow chart of the simulation process.

5 Results and Discussions

Figures 3, 4 and 5 present the simulated results for engine parameters using biodiesels from various feedstocks such as yellow grease, palm, rapeseed, soy, canola, sunflower, tallow, corn, algae, coconut, camelina, safflower, waste cooking oil, castor, and others (Peanut, karanja, rice bran, sorghum, lard, poultry fat, mustard, peanut, babarsu, fish oil, kusum, beheda and groundnut) in a stationary 4-cylinder water cooled compression ignition engine.

In Fig. 3, the simulated specific fuel consumptions (BSFC) of the biodiesel from the stated feedstock were displayed. The least specific fuel consumption was exhibited by safflower, followed by palm and then corn. The highest BSFC was depicted by castor followed by coconut and then tallow. While the BSFC of safflower was recorded as 0.335 kg/kW-h, the BSFC of castor was 0.614 kg/kw-h. The BSFC values were in line with the outcomes of Chalatlon et al. [22], Enweremadu and Mbarawa [23], Ong et al. [24], Mofijur et al. [25] and Rahim et al. [26]. Parameters like the volumetric fuel consumption, kinematic viscosity, cetane number, density of the fuel and calorific value of the biodiesel have major influence on the brake specific fuel consumption. At this engine speed of 2500 rpm, the fuel consumptions were 0.449, 0.373, 0.474, 0.450, 0.0.491, 0.436, 0.594, 0.394, 0.498, 0.602, 0.518, 0.335, 0.482 and 0.614 kg/kW-h for yellow grease, palm, rapeseed, soy, canola, sun flower, tallow,

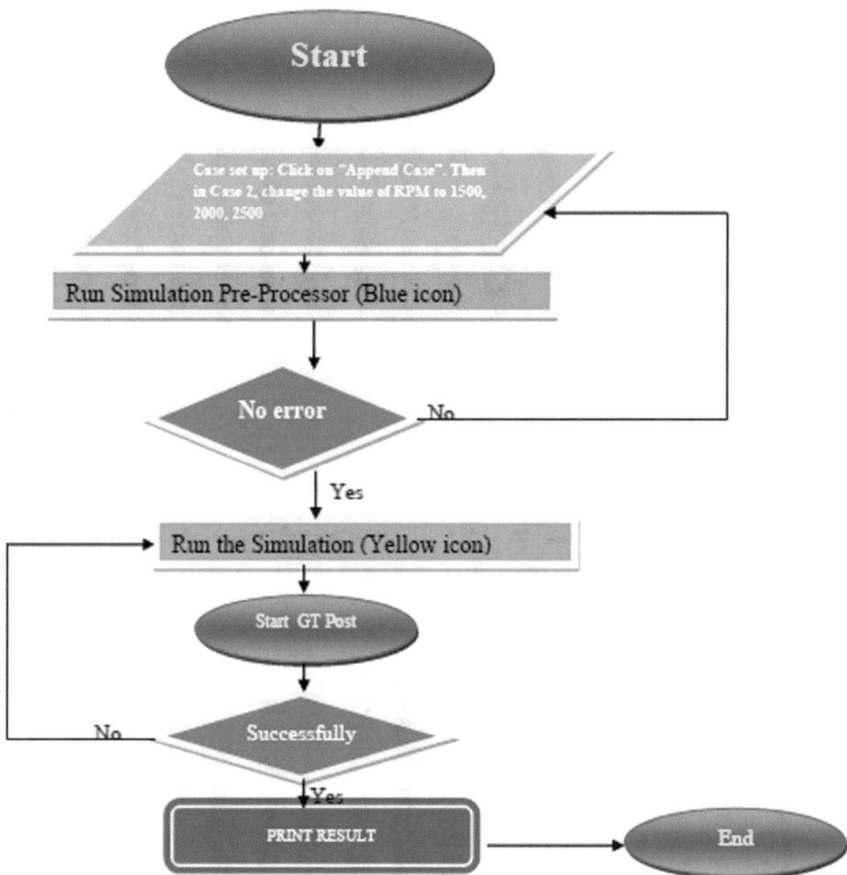

Fig. 2 Simple flowchart for the execution of GT-Power program

corn, algae, coconut, camelina, safflower, waste cooking oil and castor respectively. The least value of BSFC was portrayed by the safflower biodiesel which was in line with works of Enweremadu et al. [23].

The BMEP is the average pressure the measure brake power output has generated, when the pistons uniformly imposed from the top death centre to the bottom death centre during each power stroke. Accordingly, the simulated results of brake mean effective pressures (BMEP) for various biodiesel feedstocks were presented in Fig. 4. Their corresponding values are 5125, 5220, 5266, 6106, 4581, 6199, 6158, 5420, 5448, 5228, 5312, 5685, 6101, and 4538 MPa for yellow grease, palm, rapeseed, soy, canola, sun flower, tallow, corn, algae, coconut, camelina, safflower, waste cooking oil and castor respectively. Castor had the least BMEP followed by canola, while the highest BMEP was observed during the simulation of the CI engine performance using sun flower, tallow, waste cooking oil and soy biodiesels respectively. The

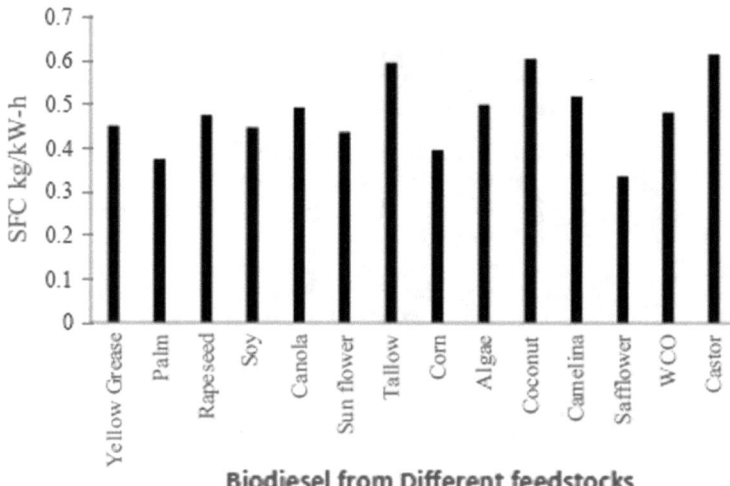

Fig. 3 Simulated specific fuel consumption of different biodiesel feedstocks

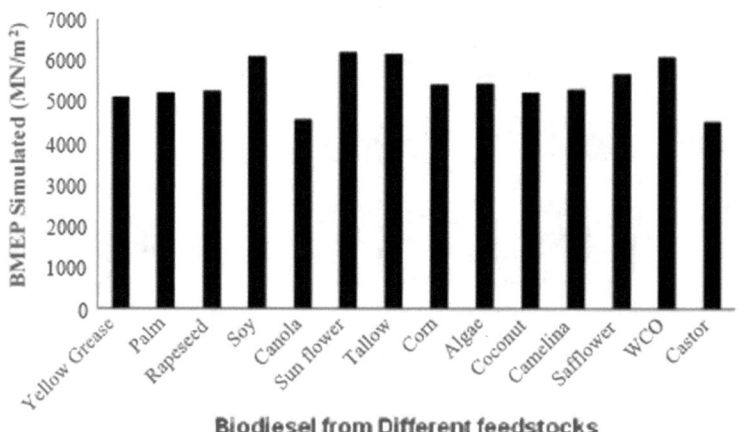

Fig. 4 Simulated brake mean effective pressure of different biodiesel feedstocks

BMEP values obtained during the simulation were in close range with the results of Rahim et al. [27] and Mohiuddin et al. [28].

The BTE is the ratio of brake power to the input energy of the fuel. The simulated results for brake thermal efficiencies (BTE) of yellow grease, palm, rapeseed, soy, canola, sun flower, tallow, corn, algae, coconut, camelina, safflower, waste cooking oil and castor were presented in Fig. 5. Their corresponding values were 0.55, 0.65, 0.51, 0.70, 0.46, 0.69, 0.57, 0.62, 0.55, 0.43, 0.57, 0.64 and 0.59% respectively. The brake thermal efficiency values were consistent with the findings of Kaisan and Pam [29] and Rahim et al. [27]. It was perceived that the consumption of biodiesel fuel

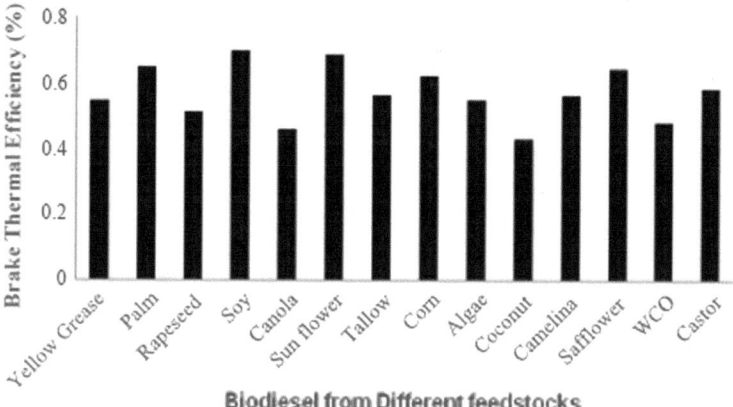

Fig. 5 Simulated brake thermal efficiencies of different biodiesel feedstocks

which is rich in oxygen supports a better formation of mixture as well as combustion process, thus enhancing the brake thermal efficiencies. According to Ong et al. [24], higher brake thermal efficiency gives engine a smoother operation, this boosting the atomization of the fuel and leading to enhanced air-fuel intermixing. Because the cetane number of biodiesel is higher than that of the pure diesel, biodiesel provides a good ignition quality. The highest brake thermal efficiency (BTE) values were given by soy, sun flower, palm and safflower respectively during the simulations with GT-Power. This implies that these four biodiesel blends can be used efficiently in diesel engine. However, the least values of brake thermal efficiency were that of coconut and waste cooking oil, although their values were also acceptable and agree with the findings of Ong et al. [24]. From the results of exhaust gas analysis, the values of Carbon monoxide were 0.1, 0.1, 0.08, 0.1, 0.2, 0.03, 0.1, 0.2, 0.1, 0.9, 0.08, 0.09, 0.03, 0.12, 0.9, 0.03 g/m^3 for case of yellow grease, palm, rape seed, soy, jatropha, tallow, corn, algae, coconut, camelina, safflower, waste cooking oil, castor and cotton respectively. The minimum concentration of CO was 0.03 g/m^3 as depicted by Sunflower and safflower based biodiesel. Cotton based biodiesel had value of 0.04 g/m^3 which remained same for all three speeds of 1000, 1500, 2000 and 2500 rpm. The maximum values for CO emission was 0.9 g/m^3 as portrayed by algae biodiesel, this is consistent with the findings of Ashok et al. [30] and Sani et al. [31]. Algae is one of the main sources of carbon dioxide, this is the likely reason that accounts for the high CO emission in algae biodiesel, because, in micro algae plant, there is CO emission which is easily converted to CO_2 via the oxidation reaction from photosynthesis of the plant. That could be the main reason why micro algae are regarded as a great source of CO_2. Similarly, the emissions of oxides of Nitrogen for yellow grease, palm, rape seed, soy, jatropha, tallow, corn, algae, coconut, camelina, safflower, waste cooking oil, castor and cotton were investigated. The least value was 0.11 g/m^3, depicted by sunflower, followed by Safflower with 0.14 g/m^3, and the peak values were 0.37 g/m^3 for Soy, followed by waste cooking oil (WCO) 0.28 g/m^3. Soy is a leguminous crop with Nitrogen fixing tendency, this accounts for the reason for its highest NO_x emissions. Generally, the amount of CO and NO_x produced during the simulation of biodiesel from different feedstock is not too much as compared to that of pure diesel case (0.36 g/m^3) which is only close to the value of Soy biodiesel emission.

6   Conclusions

Biodiesel is a clean fuel with lower emissions and can be used to replace conventional diesel wholly or partially in a compression ignition engines based on the results aforementioned, this is in concordance with the findings of Kaisan et al. [32–43], Sani et al. [44] and Dandaje and Kaisan [45]. From the simulated results obtained, all the feedstocks reviewed conform to ASTM and EN Standards, except castor for viscosity and heating value; algae for density and specific gravity; and palm and tallow for pour point. This implies that, the quality of atomization, combustion, fuel droplets and air-fuel mixing can be improved in CI engines using these blends Therefore, the fuel properties indicate that they have the potential to refill the partial energy demands in an eco-friendly way and its likelihood to be a viable fuel source for CI engines. Biodiesel yields extraordinary emissions chattels and produces a lesser amount of greenhouse gases emissions as compared with diesel. Pure biodiesel from cotton, jatropha and neem gave the lower CO and NO_x emissions compared with the petro diesel. Cotton biodiesel had the same value of CO emissions which was 0.04 g/m^3 throughout the experiment at all engine speed as well as during simulation, this is in line with the findings of Kaisan et al. [46]. The least values of CO emission were 0.03 g/m^3 both for Sunflower and Safflower biodiesel which also had the lease values of NO_x emissions as 0.11 and 0.14 g/m^3. The highest BSFC was depicted by castor followed by coconut and then tallow, the highest bmep was observed during the simulation of the CI engine performance with sun flower, tallow, waste cooking oil and soy biodiesels in that order. The highest brake thermal efficiency values were given by soy, sun flower, palm and safflower respectively during the Cussons stationary 4-cylinder CI engine simulations with GT-Power software.

References

1. Banks, J., Carson, J., Nelson, B., & Nicol, D. (2001). *Discrete-event system simulation* (Vol. 1, Issue 3, pp. 1–3). Prentice Hall.
2. Bello, R. A. (2014). Model building and usage in social science research. In *Leading issues in general studies* (Vol. 1, pp. 1–5). University of Ilorin.
3. Kolade, B., Morel, T., & Kong S. (2014). *Coupled 1-D/3-D analyses of fuel injection and diesel combustion*. Retrieved on October 9, 2014 from https://www.erc.wisc.edu/documents/13-GT-KIVA-Modeling-Meeting.pdf.

4. Bhave, A., Kraft, M., Montorsi, L., & Fabian, M. (2004). *Modelling a dual fuelled multi cylinder HCCI engine using a PDF based engine cycle simulator* (pp. 1–14). SAEInternational 01(0561).
5. Rahim, R., Mamat, R., & Taib, M. Y. (2012). Comparative study of biofuels and biodiesel blend with mineral diesel using one dimensional simulation. *Material Science and Engineering, 36*(012009), 1–10.
6. Rahim, R., Mamat, R., Taib, M. Y., & Abdullah, A. A. (2012). Influence of fuel temperature on a diesel engine performance operating with biodiesel blend. *International Journal of Advance Sciences and Technology., 43,* 115–126.
7. Kakee, A. H., & Pishgoogie, M. (2011). Determination of optimal valve timing for internal combustion engines using parameter estimation method. *International Journal of Automotive Engineering., 1*(2), 130–140.
8. Ren, Y., Abu-Ramadan, E., & Li, X. (2010). Numerical simulation of biodiesel fuel combustion and emission characteristics in a direct injection diesel engine. *Frontiers of Energy and Power in China, 4*(2), 252–261.
9. Graciano, V., et al. (2016). Modeling and simulation of diesel, biodiesel and biogas mixtures driven compression ignition internal combustion engines. *Energy research, 40*(1), 100–111.
10. Aldhaidhawi, M., et al. (2016). *IOP Conference Series: Materials Science and Engineering, 147,* 012135.
11. Raut, L. P. (2013). Computer simulation of CI engine for diesel and biodiesel blends. *International Journal of Innovative Technology and Exploring Engineering, 3*(2).
12. Zheng, Junnian. (2009). *Use of an engine cycle simulation to study a biodiesel fueled engine.* Mscthesis: Texas AM University.
13. Morse, K. D. (2014). *Numerical analysis of biodiesel combustion in a direct injection compression ignition engine.* Georgia Southern University Electronic Thesis & Dissertations, Paper 2185.
14. Boggavarapu, P., & Ravikrishna, R. V. (2013). A review on atomization and sprays of biofuels for IC engine applications. *International Journal of Spray and Combustion Dynamics, 5*(2), 85–121.
15. Nasim, N., Yarasu, B. R., & Yamin, J. J. (2010). Simulation of CI engine powered by neat vegetable oil under variable fuel inlet temperature. *Indian Journal of Science and Technology, 3*(4), 87–92.
16. Harch, C. A., Rasul, M. G., Hassan, N. M. S, & Bhuiya, M. M. K. (2014). Modelling of engine performance fuelled with second generation biodiesel. In *10th International Conference on Mechanical Engineering, ICME2013, Procedia Engineering* (Vol. 90, pp. 459–465).
17. Al-Dawody, Mohamed F., & Bhatti, S. K. (2013). Theoretical modelling of combustion characteristics and performance parameters of biodiesel in DI diesel engine with variable compression ratio. *International journal of energy and environment, 4*(2), 231–242.
18. Erinc, et al. (2013). A brief review on examination of experimental and numerical studies about using of biodiesel on diesel engines. *Journal of naval science and engineering, 9*(2), 33–49.
19. Hu, Q., Sommerfeld, M., Jarvis, E., Ghirardi, M., Posewitz, M., & Seibert, M. (2008). Micro Algal Tri-acyl-glycerol as feedstock for biofuel production: Perspectives and advances. *Plant Journal, 54,* 621–639.
20. Kaisan, M. U. (2017). *Modelling and simulation of biodiesel blends into compression ignition engine.* Unpublished Ph.D. Research, Ahmadu Bello University, Zaria, Nigeria.
21. Kaisan, M. U. (2014). *Determination of physico-chemical properties of biodiesel from wild grape seeds/diesel blends and their effects on the performance of a diesel engine.* Unpublished M.Sc. Research, Ahmadu Bello University, Zaria, Nigeria.
22. Chalatlon, V., Roy, M. M., Dutta, A., & Kumar, S. (2011). Jatropha oil production and experimental investigation of its use as an alternative fuel in a DI diesel engine. *Journal of Petroleum Technology and Alternative Fuels, 2*(5), 76–85.
23. Enweremadu, C. C., & Mbarawa, M. M. (2009). Technical aspects of production and analysis of biodiesel from used cooking oil-A review. *Renewable & Sustainable Energy Reviews, 13*(9), 2205–2224.

24. Ong, H. C., Masjuki, H. H., Mahlia, T. M. I., Silitonga, A. S., Chong, W. T., & Leong, K. Y. (2014). Optimization of biodiesel production and engine performance from high free fatty acid Calophylluminophyllum oil in CI diesel engine. *Energy conversion and Management, 81,* 30–40.

25. Mofijur, M., Masjuki, H. H., Kalam, M. A., Atabani, A. E., Rizwanal Fattah, I. M., & Mobarak, H. M. (2014). Comparative evaluation of performance and emission characteristics of Moringa Oleifera and palm oil based biodiesel in a diesel engine. *Journal of Industrial Crops and Product., 53,* 78–84.

26. Rahim, R., Mamat, R., & Taib, M. Y. (2012). Comparative study of biofuels and biodiesel blend with mineral diesel using one dimensional simulation. *Material Science and Engineering, 36*(012009), 1–10.

27. Rahim, R., Mamat, R., Taib, M. Y., & Abdullah, A. A. (2012). Influence of fuel temperature on a diesel engine performance operating with biodiesel blend. *International Journal of Advance Sciences and Technology, 43,* 115–126.

28. Mohiuddin, A. K., Ataur-Rahman, M., & Dzaidin, M. (2007). Optimization design of automobile exhaust system using GT-power. *International Journal of Mechanical and Material Engineering, 2*(1), 40–47.

29. Kaisan, M. U., Pam, G. Y., & Kulla, D. M. (2013). Physico-chemical properties of biodiesel from wild grape seeds oil and petro-diesel blends. *American Journal of Engineering Research, 02*(10), 291–297.

30. Ashok, B. Nanthagopal, K., Kavalipurapu, R. T., Saravanan, B. D., Arumuga P., & Kaisan M. U. (2018). Experimental study of methyl tert-butyl ether as an oxygenated additive in diesel and calophyllum inophyllum methyl ester blended fuel in CI engine. *Environmental Science and Pollution Research.* https://doi.org/10.1007/s11356-018-3318-y. Published by Springer.

31. Sani, S., Kaisan M. U., Kulla, D. M. Ashok, B., & Nanthagopal, K. (2018). Determination of physico chemical properties of biodiesel from Citrullus lanatus seeds oil and diesel blends. *Industrial Crops and Products, 122,* 702–708. https://doi.org/10.1016/j.indcrop.2018.06.002. Published by Elsevier.

32. Kaisan, M. U., Anafi, F. O., Nuszkwoski, J., Kulla, D. M., & Umaru, S. (2016a). GC-MS analyses of biodiesel produced from cotton seed oil. *Nigerian Journal of Solar Energy, 27,* 56–61. Published by Solar Energy Society of Nigeria.

33. Kaisan, M. U., Anafi, F. O., Nuszkwoski, J., Kulla, D. M., & Umaru, S. (2017a). Exhaust emissions of biodiesel binary and multi-blends from cotton, jatropha and neem oil from stationary multi cylinder CI engine. *Transportation Research Part D: Transport and Environment, 53,* 403–414. Published by Elsevier.

34. Kaisan, M. U. Anafi, F. O. Nuszkwoski, J. Kulla, D. M., & Umaru, S. (2017b). Fuel properties of biodiesel from cotton, jatropha binary and multi-blends with diesel. *Nigerian Journal of Material Science and Engineering, 7*(1), 55–61. Published by Material Science and Technology Society of Nigeria.

35. Kaisan, M. U., Anafi, F. O., Nuszkwoski, J., Kulla, D. M., & Umaru, S. (2017c). Cold flow properties and viscosity of biodiesel from cotton, jatropha and neem binary and multi-blends with diesel. *African Journal of Renewable and Alternative Energy, 1*(1), 80–89. Published by Renewable and Alternative Energy Society of Nigeria.

36. Kaisan, M. U., Anafi, F. O., Nuszkowski, J., Kulla, D. M., & Umaru, S. (2017d). Calorific value, flash point and cetane number of biodiesel from cotton, jatropha and neem oil binary and multi-blends with diesel. *Biofuels, 8*(4), 1–7. Published by Taylor and Francis.

37. Kaisan, M. U., Anafi, F. O., Nuszkowski, J., Kulla, D. M., & Umaru, S. (2017e). Performance evaluation of binary and multi-blends of biodiesel from cotton, jatropha and neem oils in multi cylinder CI engine. *African Journal of Renewable and Alternative Energy, 2*(1). Published by Renewable and Alternative Energy Society of Nigeria.

38. Kaisan, M. U., Nafiu, T., & Habib, Y. B. (2010). Carbon capture storage and processing as a means of enhancing renewable energy sources in Nigeria. *Renewable and Alternative Energy for Sustainable National Development, 2*(1), 219–225.

39. Kaisan, M. U., Naifu, T., & Habib, Y. B. (2010). Towards new policies in minimizing green house gas emission in Nigeria. *Renewable and Alternative Energy for Sustainable National Development, 2*(1), 306–316.
40. Kaisan, M. U., Abubakar, S., Umaru, S., Dhinesh, B., Mohamed Shameer, P., & Mohamed Nishath, P. (2018). Comparative analyses of biodiesel produced from jatropha and neem seeds oil using gas chromatographic and mass spectroscopic technique. *Biofuels*, 1–12. https://doi.org/10.1080/17597269.2018.1537206.
41. Kaisan, M. U., Anafi, F. O., Nuszkwoski, J., Kulla, D. M., & Umaru, S. (2016b). *Review of physico-chemical properties of biodiesel feedstock and their suitability for nigerian economy.* Nigerian Energy Forum, Oriental Hotel, Lagos, Nigeria.
42. Kaisan, M. U., Anafi, F. O., Nuszkwoski, J., Kulla, D. M., & Umaru, S. (2016c). Fuel properties of biodiesel from cotton, jatropha binary and multi-blends with diesel. Material Science and technology Society of Nigeria. NARICT, July, 2016.
43. Kaisan, M. U., Anafi, F. O., Nuszkwoski, J., Kulla, D. M., & Umaru, S. (2016d). Performance evaluation of binary and multi-blends of biodiesel from cotton, jatropha and neem oils in multi cylinder CI engine. In *International conference on alternative fuels*, University of Kayseri, Turkey.
44. Sunusi, S., Kulla, D. M., Kaisan, M. U., & Liman, B. M. (2017). Production of *Citrullus Lanatus* derived biodiesel and gas chromatography mass spectroscopic analyses. *African Journal of Renewable and Alternative Energy, 2*(3). Published by Renewable and Alternative Energy Society of Nigeria.
45. Dandajeh, H. A., & Kaisan M. U. (2014). Engine performance characteristics of a gardener compression ignition engine using rapeseed methyl ester. In *Nigeria engineering conference*, Ahmadu Bello University, Zaria, Nigeria.
46. Kaisan, M. U., Anafi, F. O., Nuszkowski, J., Kulla, D. M., & Umaru, S. (2017f). Embellishments of blend ratio of binary and multi- blends of biodiesel from cotton jatropha and neem oils on CI engine performance and emissions at various speeds. In *Conference of automotive engineers institute* (Vol. 7, pp. 57–69), National Engineering Centre, Abuja, Nigeria.

Agricultural Wastes as Feedstock for Thermo-Chemical Conversion: Products Distribution and Characterization

Samarjit Gogoi, Nilutpal Bhuyan, Debashis Sut, Rumi Narzari, Lina Gogoi and Rupam Kataki

Abstract Pyrolysis is one of the most promising thermal conversion technologies for biomass conversion. In pyrolysis, biomass is decomposed into solid char, bio-oil, and gas, that satisfies different forms of energy requirement. Further, Pyrolysis also offers the unique advantage of utilizing diverse lingo-cellulosic biomasses including agricultural wastes as feedstocks. Globally, India is the second largest sesame seed producing country. So, a vast amount of sesame stalk is generated as an agricultural waste in India. In this work, slow pyrolysis of sesame stalk was done at four different temperatures in the range of 350–650 °C at a heating rate of 40 °C/min in a laboratory-scale fixed bed reactor. The variation of pyrolysis products yield with temperature was observed, and all the products were quantified. The liquid and solid products of pyrolysis i.e. bio-oil and biochar respectively, were characterized by elemental analyser (CHN analyser), GC-MS, SEM and FTIR. From the study it was observed that pyrolysis temperature had significantly affected the yield of pyrolysis products. Biochar yield decreased while volatile gas yield increased with increase in temperature. The bio-oil yield was maximum at 550 °C and it was found that the higher heating value (HHV) of the oil was 26.89 MJ/kg. The carbon content of the biochar also increased with the temperature and hence the HHV. The pH of the biochar increased with the temperature and basic in nature which is suitable for soil

S. Gogoi · N. Bhuyan · D. Sut · R. Narzari · L. Gogoi · R. Kataki (✉)
Biofuel Laboratory, Department of Energy, Tezpur University, Tezpur 784028, Assam, India
e-mail: rupamkataki@gmail.com

S. Gogoi
e-mail: samarjitipl123@gmail.com

N. Bhuyan
e-mail: nilutpal6bhuyan@gmail.com

D. Sut
e-mail: debashissut07@gmail.com

R. Narzari
e-mail: pinki.lkt@gmail.com

L. Gogoi
e-mail: lina.dbr@gmail.com

© Springer Nature Singapore Pte Ltd. 2020
S. K. Ghosh (ed.), *Energy Recovery Processes from Wastes*,
https://doi.org/10.1007/978-981-32-9228-4_10

amendment in north-eastern region. The study established that the sesame stalk was a suitable raw-material for pyrolytic valorization.

Keywords Thermo-chemical conversion · Sesame stalk · Agricultural waste · Biochar · Bio-oil

1 Introduction

The continuous exhaustion of petroleum reservoirs, rising fuel demand, and the change in the environment due to the burning of petroleum fuel has augmented the interest on the alternative energy sources. The renewable sources could overcome the above issues due to their renewability, biodegradability, environmentally benign nature, extensive accessibility, and cost effectiveness. Biomass is considered as one of the most favourable, clean and sustainable alternative energy source among all renewable sources. Currently, thermo-chemical conversions (pyrolysis, gasification and combustion) of biomass into biofuels and other products have received special attentions [1]. In this regard, a thorough understanding of the thermal behaviour of the feedstock as well as fuel characteristics are necessary for planning of thermo-chemical conversion processes like pyrolyzers, gasifiers, etc. [2]. Pyrolysis is an attractive, convenient, economical and environmentally suitable technology in which all types of biomass including residues obtained from agricultural sectors, forests can be thermally decomposed under inert atmosphere to produce syngas, bio-oil and biochar [2, 3]. In the biomass pyrolysis process, various reactions such as dehydration, depolymerization, repolymerization, condensation etc. are taking place concurrently. Bio-oil can be used as a fuel which is favourable for storing and in transport sector. Also, various chemicals can be extracted from bio-oil due to the presence of hundred types of organic compounds like carboxylic acids, alcohols, aldehydes, phenols etc. [3]. Due to the high calorific value compared to its feedstock bio-oil is preferred for mixing with traditional diesel. It can be mixed with the conventional diesel because of its higher calorific value than the parent feedstock [4]. Also it can be used in furnaces, boiler for heat and power generation, diesel engines and kilns [5]. Pyrolysis of biomass mostly produces hydrogen, methane, carbon monoxide, carbon dioxide and other lighter hydrocarbons as gaseous products, which has the potential to substitute natural gas. Biochar is the solid product obtained from pyrolysis process. It can be used as fuel, as a soil amendment, carbon sequester and as an activated carbon owing to its high porosity and surface area [6].

The pyrolysis behaviour of various agricultural wastes such as sawdust, rice husk, straw, tobacco waste, red pepper waste, corncob, oreganum stalks and mangaba seed. has been already explored [1, 3, 7, 8]. Unchecked stubble/agricultural wastes burning is a prime cause for environmental degradation. Moreover, these residues have little or no market value. So, pyrolysis can be a viable option to resolve these problems by producing various value added products [5].

Sesame (*Sesamum indicum L.*) is considered as one of the oldest cultivated oilseed crops in the world. India ranks second both in area and production of sesame and hence a vast amount of sesame stalks are being generated as an agricultural waste [9]. The major sesame seeds producing states are Maharashtra, Rajasthan, West Bengal, Andhra Pradesh, Gujarat, Tamil Nadu, Madhya Pradesh, and Telangana. India produces 1.38 MMT of sesame stalks annually [10].

The present investigation aimed at pyrolytic conversion of sesame stalk in a lab-scale reactor (fixed bed) to find the operating temperature for maximum liquid yield. Also, the impact of various temperatures on the yield of product was investigated. In this work, slow pyrolysis of sesame stalk was carried out at four different temperatures ranging from 350 °C to 650 °C at a heating rate of 40 °C/min. The liquid and solid products from pyrolysis are characterized by using various techniques.

2   Materials and Methods

2.1   Sample Preparation and Analysis

Sesame stalk was collected from the KVK-Tezpur, Sonitpur district, Assam, India, and sun dried to reduce the residual moisture content. These sesame stalks were then grinded and passed through 0.42 mm sieve. The resultant samples were dried in an oven at 100 ± 5 °C for 6 h before experiment (Fig. 1).

The lignocellulosic contents of sesame stalk, i.e. extractive contents, cellulose, hemicellulose and lignin, were determined according to the methods described in Ayeni et al. [11].

The proximate analysis of sesame stalk were done as per ASTM Standards [ASTM D3174 (ash), ASTM D3173 (moisture) and ASTM D3175 (volatile matter)]. Fixed carbon percentage in the samples was calculated by first adding percentage content of moisture, volatile matter and ash, followed by subtracting this sum from 100.

(a)   **(b)**

Fig. 1 a Sesame stalk and **b** grinded sesame stalk

Thermo Finnigan, Flash EA 1112 was used to carry out the ultimate analysis. Calorific value was determined by ASTM D2015 method, using an auto bomb calorimeter (model: 5E-1AC/ML). The structure of the sample was investigated on a Nicolet IR spectrometer at room temperature in the spectral range of 4000–400 cm^{-1}. Thermal behaviour of the biomass was investigated in a thermo-gravimetric (TGA) instrument (Perkin Elmer STA 6000).

2.2   Experimental Set up

To carry out the pyrolysis of various samples, a reactor–furnace system coupled with PID controller was used. In the current investigation 10 g of sample was subjected to pyrolysis with a heating rate of 40 °C/min at desired temperature from 350 to 650 °C in nitrogen atmosphere. To collect the condensable gas during the process, a condenser was attached to the setup. Using Diethyl ether as a solvent the liquid product was recovered. The aqueous portion of the liquid product was separated from the organic portion using a separating funnel. The solid portion left as residue was evaluated as biochar. The non-condensable gas yield was determined from the difference (Fig. 2).

2.2.1   Products Characterization

Only the liquid product obtained at operating temperature at which liquid yield was maximum, was subjected to characterization. The chars obtained at four different temperatures in the range of 350–650 °C were analysed. For proximate and ultimate analysis as well as chemical composition the experiments mentioned for biomass characterization were used.

A pH meter (model: EUTECH pH 510) and a digital TDS/conductivity meter (model: MK 509) was used to determine the pH and electrical conductivity (EC) of

Fig. 2 Pyrolysis set-up

biochar by the methodology described in detail elsewhere [12]. Scanning electron microscope (SEM) analyses of biochar were performed on a Jeol, JSM-6290LV instrument. GC-MS experiment of bio-oil was conducted in Perkin Elmer Claurus 600 instrument coupled with TCD detector.

3 Results and Discussion

3.1 Characteristics of Biomass

From the proximate analysis, the moisture content of sesame stalk was found to be 7.98% which made sesame stalk a good option to use for thermo-chemical conversion [13]. The volatile matter content of the sesame stalk was high (70.72%) and ash content was low (11.24%), which indicated a low ignition temperature and hence ideal for application in gasifier and pyrolysis. The presence of higher volatile matter in the feedstock was desirable because it made the sample more reactive than solid fuels with lower volatile content as the sample more readily devolatilized leaving less amount of fixed carbon content. Sesame stalk also had lower fixed carbon content (10.06%) than coal which also supported that calorific values of sesame stalk (15.57%) was lower than that of coal [13] (Table 1).

The lower amount of nitrogen (1.33%) in the feedstock indicated that NO_x emissions from sesame stalk was not much of environmental concern than the combustion of fossil fuels. The H/C and O/C ratios for sesame stalk were 2.51 and 0.75, respectively. Calorific value of sesame stalk was determined to evaluate its energy content which is found to be 15.57 MJ/Kg. From the biochemical analysis it was observed that sesame stalk contained 31.17% wt. of lignin which gave higher char yield than cellulose and hemicelluloses (Table 2).

Table 1 Physico-chemical analysis of biomass

Properties	Sesame stalk
Moisture content (%)	7.98
Ash content (%)	11.24
Volatile matter (%)	70.72
Fixed carbon (%)	10.06
C (%)	44.65
H (%)	9.34
N (%)	1.33
O (%)	44.68
H/C	2.51
O/C	0.75
Calorific value (MJ/kg)	15.57

Table 2 Bio-chemical analysis of biomass

Bio-chemical analysis			
Extractives (%)	Cellulose (%)	Hemicelluloses (%)	Lignin (%)
18.29	32.34	18.20	31.17

Fig. 3 FTIR spectra of biomass

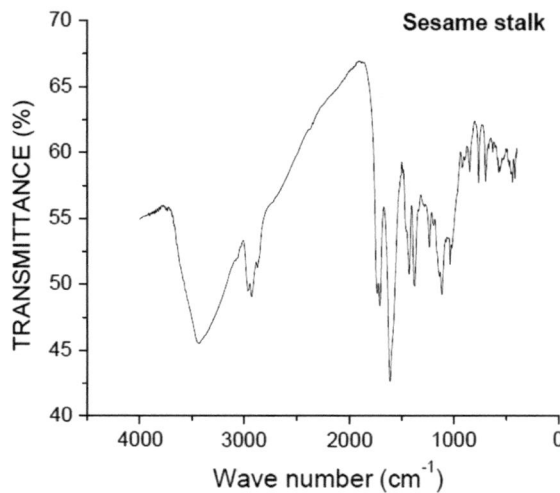

3.2 FTIR Spectra of Biomass

FTIR spectra and band assignments of sesame stalk were presented in Fig. 3. The stretching vibration of bonded and free –OH groups was attributed by 3470 cm^{-1}. The band appeared at 2922 cm^{-1} was caused by C–H stretching vibrations of methylene group. The band at 1730 cm^{-1} proves the presence of hemicelluloses, indicated by the C=O stretching vibration of non-conjugated ketones, carbonyls and ester groups. The sharp peak obtained at 1618 cm^{-1} attributed to the C=C stretching vibrations of aromatic rings. The fingerprint region of the spectrum from 1500 to 1200 cm^{-1} was dominated by C–H, N–O and O–H vibrations. The small absorbance at 890 and 780 cm^{-1} corresponded to alcohol (primary and secondary) and aliphatic ester, C–O–C, C–O stretching and C–H bend of aromatic ring. On the other hand two bend at 1060 and 620 cm^{-1} were related to the C–OH stretching vibration of the cellulose and hemicelluloses and C–H vibration of aromatic rings.

3.3 Thermal Analysis

The thermal decomposition of sesame stalk was carried out at heating rate of 20 °C/min with nitrogen flow rate of 50 ml/min and their derivative curves (DTG),

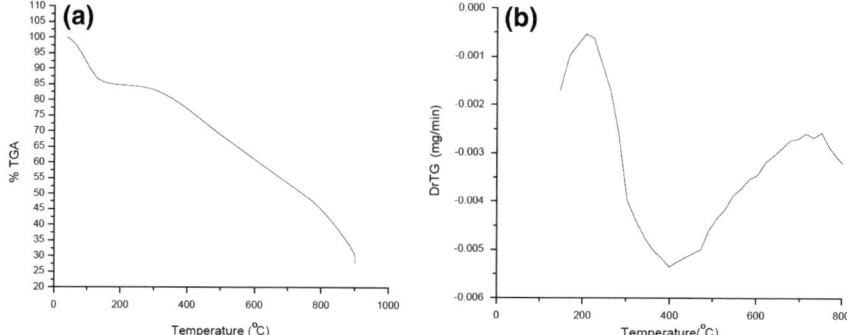

Fig. 4 **a** TGA and **b** DTG of sesame stalk

were shown by TG patterns as presented in Fig. 4a, b. The mass loss curves of these samples exhibited three main weight loss regions, which corresponded to devolatilization and solid decomposition. The initial weight loss of 7.98 wt% occurred in the temperature range of 30–200 °C was due to the evaporation of moisture content of sesame stalk, which was known as dehydration zone. The devolatilization step of sesame stalk proceeded in a range from approximately 200–310 °C might be due to the release of volatile matter as a result of breaking of weaker chemical bonds. The hemicellulose and cellulose decomposition occurred in this region, while lignin was decomposed throughout the temperature range of the pyrolysis process without showing any characteristics DTG peaks [14].

3.4 Pyrolysis Products Yield

Pyrolysis of sesame stalk was done at a constant heating rate of 40 °C/min at four different temperatures viz. 350, 450, 550 and 650 °C to study the variation in products yield with pyrolysis temperature.

From Fig. 5, it was observed that the yield of biochar decreased from 34.45 to 28.72% with the temperature of the pyrolysis experiment increasing from 350 to 650 °C. The reduction in the char production was observed with raise in temperature owing to the greater primary decomposition of biomass into low molecular weight organics and gases as well as the occurrence of secondary decomposition of char residues at elevated temperature [15].

Bio-oil yield also increased from 19.76 to 24.85% in the temperature range of 350–550 °C. However with the increase in pyrolysis temperature to 650 °C, decrease in bio-oil yield (22.34%) was observed. The reason for this trend was due to the secondary cracking reaction resulting the production of more non-condensable low molecular weight gases at higher temperature. The yield of gas increased from 17.45 to 24.46% with increasing pyrolysis temperature which was also due to the secondary cracking of pyrolysis vapours at higher temperatures [16].

Fig. 5 Distribution of pyrolysis product yield with temperature

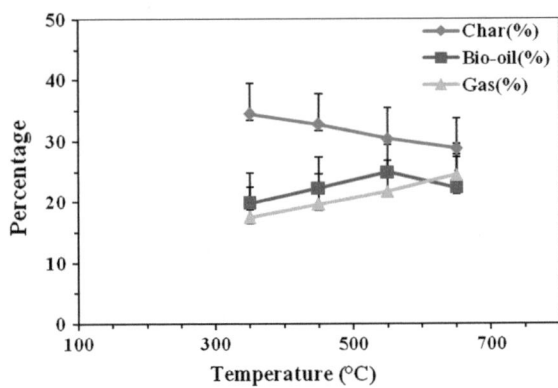

3.5 Characterization of Biochar

As observed in Table 3, volatile matter content of biochar decreased from 33.45 to 27.12% with ascending pyrolysis temperature from 350 to 650 °C for sesame stalk. Also, the volatile matter percentage was lower in biochar compared to the parent biomass. Increase in fixed carbon content (49.34–56.20%) in biochar with the increasing temperature indicated the further cracking of the volatiles fractions into low molecular weight liquids and gases. As the temperature increased percentage of ash content also varies (increased from 6.57 to 8.20%). This was due to the gradual

Table 3 Proximate and ultimate analysis of biochar

Properties	Biochar			
	350 °C	450 °C	550 °C	650 °C
Moisture content (%)	9.43	8.58	8.20	7.28
Ash content (%)	6.57	7.65	7.90	8.20
Volatile matter (%)	33.45	30.80	29.92	27.12
Fixed carbon (%)	49.34	52.48	53.38	56.20
pH	8.98	9.28	9.43	9.58
EC (m℧)	0.135	0.143	0.225	0.401
BET surface area (m^2/g)	–	4.152	6.828	–
C (%)	67.55	69.67	70.34	72.65
H (%)	3.18	3.06	2.87	2.64
N (%)	5.6	5.1	6.6	5.26
O (%)	23.67	22.17	20.19	19.45
H/C	0.56	0.53	0.49	0.44
O/C	0.26	0.24	0.22	0.20
Calorific value (MJ/kg)	16.82	20.34	21.76	24.32

increase in concentration of inorganic constituents [17]. The moisture percentage contained in bio-char decreased with ascending temperature which indicated the extra surface hydrophobicity of bio-char at higher temperature [18].

With the increase in pyrolytic temperature there was a sharp decline in O and H content of biochar, while the content of C has increased. The enhance degree of carbonization is the main cause for this increase in C content. Dropping off the O and H content might be due to the cleavage of weaker bonds during carbonization and, as a result elimination of H_2O, H_2, hydrocarbons, CO and CO_2 gas. Increased aromaticity of carbon content in the biochar is responsible for the decrease in the H/C molar ratios as temperature raised. Additionally, O/C molar ratio changes inversely with the temperature indicated the decline of some polar functional groups that contains O on biochar surface [17]. The lower oxygen and higher carbon content of biochar also contributed to the higher calorific value of it [19].

Table 3 also shows that the pH of biochar increased with increasing pyrolysis temperature due to the progressive loss of acidic surface groups during thermal treatment [17]. The pH of the biochar for sesame stalk were in the base range, which could be regarded as suitable for soils of North-east India, which are generally acidic in nature. Biochar with elevated pH could be used as liming agent to rehabilitate the acidic soil and hence could be used for soil amendment [19].

3.5.1 FTIR—Spectra of Biochar

The FTIR spectra of biochar obtained at two pyrolysis temperatures viz. 350 and 650 °C were shown in Fig. 6. The peak obtained around 3600–3200 cm^{-1} was due to the O–H stretch, which was clearly visible for biochar at 350 °C. But the intensity of absorbance of this peak decreased at higher temperature (650 °C) due to the dehydration of cellulose and ligneous compounds. The bands appeared at 1650–1580 cm^{-1} and 1470–1310 cm^{-1} were due to alkene and alkane groups respectively, however the peaks were less observable in high temperature. This might be due to the decrease of aliphaticity of the biochar at higher temperature. The investigation further indicated

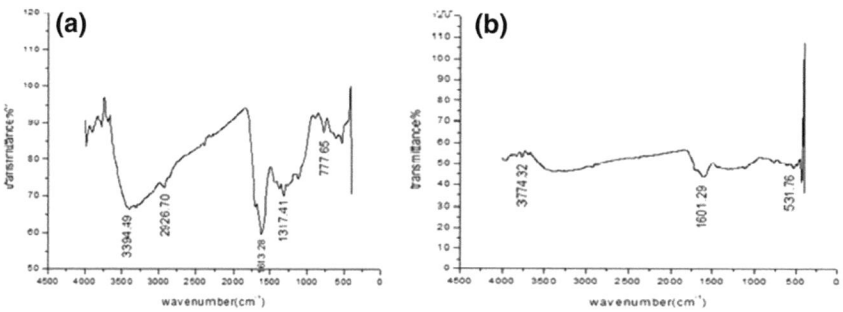

Fig. 6 FTIR spectra of biochar at **(a)** 350 °C and **(b)** 650 °C

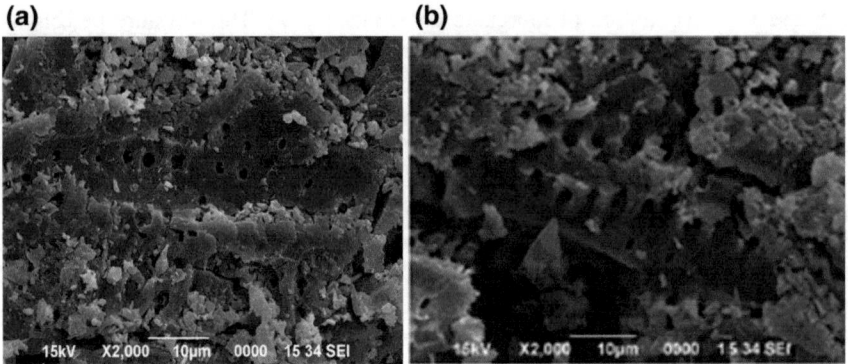

Fig. 7 SEM images of biochar at (**a**) 350 °C and (**b**) 650 °C

that the functional groups which contain hydrogen and oxygen were eliminated at high pyrolysis temperature. Thus, it was observed that with increasing temperature degree of aromaticity increased while degree of polarity decreased for biochar.

3.5.2 SEM Analysis

SEM analysis gives the information regarding the structure and morphological characteristics of a sample. SEM images of biochar obtained at 350 and 650 °C showed the porous structure of biochar (Fig. 7). These porous structures were due to pyrolysis which eliminated the volatile materials and as a result some extra adsorption sites were created. At 650 °C, the SEM images were more irregular and distorted than at low temperature (350 °C). This might be due to the development of more pores with slightly larger diameter and damage of original structure with increasing temperature. The existence of pores on the biochar provided a suitable space for microbial activity and the reservation of soil nutrients [20].

3.6 Characterization of Bio-Oil

3.6.1 Ultimate Analysis

See Table 4.

3.6.2 FTIR Spectra of Bio-Oils

The FTIR spectrum of sesame stalk pyrolysis oil was shown in Fig. 8. Stretching vibration at 3414 cm^{-1} showed the occurrence of O–H functional group of phenols

Table 4 Ultimate analysis of bio-oil

Parameters	Values
Carbon (%)	54.32
Hydrogen (%)	9.32
Nitrogen (%)	4.67
Oxygen (%)	28.32
H/C	2.06
O/C	0.39
Empirical formula	$C_{13.570}H_{27.940} N_{1.00} O_{5.306}$
Acid number	21.32
Calorific value (MJ/Kg)	26.89

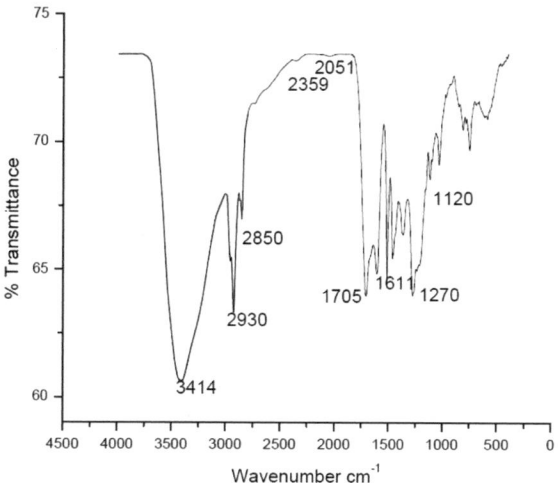

Fig. 8 FTIR spectra of bio-oil at 550 °C

and alcohols. The C–H stretching modes found between 3000 and 2800 cm^{-1} showed the presence of alkane groups. Again the presence of alkanes can be found from the C–H deformation vibrations in the range of 1350–950 cm^{-1}. Similarly C–H bending vibration between 950 and 615 cm^{-1} indicated the existence of ether as well as alcohol group. The C=O stretch appeared between 1850 and 1650 cm^{-1} were due to ketones, esters, aldehyde groups. The bands in the range of 1650 and 1580 cm^{-1} represented the C=C stretching vibrations of alkenes.

GC-MS analysis was performed to find different types of compounds in pyrolysed oil. The result indicated its complex composition and it is difficult to separate the peaks. A complete separation of all the peaks was not possible due to complicated composition of bio-oil. The total ion chromatogram (TIC) of bio-oil was shown in Fig. 9. Table 5 shows the identified components with their retention time (minutes)

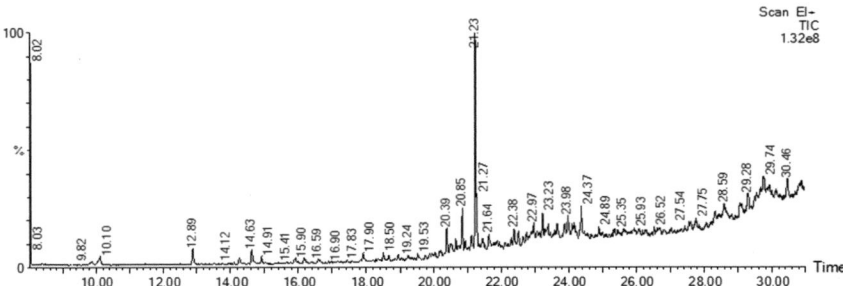

Fig. 9 Total ion chromatogram of bio-oil

Table 5 Retention time and compound name present in bio-oil

S. no.	RT (min)	Area	Compound name
1	10.10	0.44	2-Furanmethanol
2	12.90	0.79	Phenol
3	14.63	0.75	Phenol,4-methyl-
4	14.91	1.04	Phenol,2-methyl-
5	18.50	0.58	4-hydroxy-2-methylacetophenone.
6	20.39	1.23	Eugenol
7	20.85	1.97	Phenol,2-ethyl
8	21.23	7.70	Vanillin
9	25.35	0.53	Caffeine
10	29.75	3.26	Phenol,4-ethyl

and area percentage. The compounds were found to be mostly aromatics and phenols. FTIR results also supported the presence of phenols. The mostly found compounds were phenol and vanillin having peak areas around or greater than 2%.

4 Conclusion

In this study, pyrolysis of an agricultural waste i.e. sesame stalk was done at four different temperatures in a fixed bed reactor at a heating rate of 40 °C/min and the variation of products yield with the pyrolysis temperature was observed. The current study reveals that the highest amount of bio-oil was yielded at 550 °C with a calorific value of 26.89 MJ/Kg. The FT-IR analysis of bio-oil revealed that its composition was dominated by oxygenated species. The effect of pyrolytic temperature on the physico-chemical and structural characteristics of biochar was observed from the experimental results. The increase in the temperature enhanced the porosity of biochar surface, decreased the biochar yield, and increased the pH of the biochar.

The high pH of biochar makes it a suitable candidate for soil amendment. The study established the potential of Sesame stalk as a raw material for thermo-chemical conversion.

Acknowledgements Tezpur University and Guwahati Biotech Park were acknowledged by the authors for providing facilities to conduct the analysis. R. Narzari, and D. Sut also acknowledge the fellowship grants from UGC.

References

1. Quan, C., Gao, N., & Song, Q. (2016). Pyrolysis of biomass components in a TGA and a fixed-bed reactor: Thermochemical behaviors, kinetics, and product characterization. *Journal of Analytical and Applied Pyrolysis, 121,* 84–92.
2. Jenkins, B. M., Baxter, L. L., Miles, T. R. M., Jr., & Miles, T. R. M. (1998). Combustion properties of biomass. *Fuel processing technology, 54,* 17–46.
3. Saikia, R., Baruah, B., Kalita, D., Pant, K. K., Gogoi, N., & Kataki, R. (2018). Pyrolysis and kinetic analyses of a Perennial Grass (*Saccharum ravannae* L.) from north-east India: Optimization through response surface methodology and product characterization. *Bioresource Technology, 253,* 304–314.
4. Jiang, X., & Ellis, N. (2010). Upgrading bio-oil through emulsification with biodiesel: Thermal stability. *Energy & Fuels, 24,* 2699–2706.
5. Hawash, S. I., Farah, J. Y., & El-Diwani, G. (2017). Pyrolysis of agriculture wastes for bio-oil and char production. *Journal of Analytical and Applied Pyrolysis, 124,* 369–372.
6. Kumar, P. (2017). Saw dust pyrolysis: Effect of temperature and catalysts. *Fuel, 199,* 339–345.
7. Demirbas, A., Pehlivan, E., & Altun, T. (2006). Potential evolution of Turkish agricultural residues as bio-gas, bio-char and bio-oil sources. *International Journal of Hydrogen Energy, 31,* 613–620.
8. Santos, R. M., Santos, A. O., Sussuchi, E. M., Nascimento, J. S., Lima, Á. S., & Freitas, L. S. (2015). Pyrolysis of mangaba seed: Production and characterization of bio-oil. *Bioresource Technology, 196,* 43–48.
9. Karuppaiah, V., & Nadarajan, L. (2013). Host plant resistance against sesame leaf webber and capsule borer *Antigastracatalaunalis Duponchel* (Pyraustidae: Lepidoptera). *African Journal of Agricultural Research, 8*(37), 4674–4680.
10. Sukumaran, R. K., Mathew, A. K., Kumar, M. K., Abraham, A., Chistopher, M., Sankar, M. (2017). First- and second-generation ethanol in India: A comprehensive overview on feedstock availability, composition, and potential conversion yields. In *Sustainable Biofuels Development in India*, Springer.
11. Ayeni, A. O., Adeeyo, O. A., Oresegun, O. M., & Oladimeji, T. E. (2015). Compositional analysis of lignocellulosic materials: Evaluation of an economically viable method suitable for woody and non-woody biomass. *AJER, 4*(4), 14–19.
12. Bordoloi, N., Narzari, R., Chutia, R. S., Bhaskar, T., & Kataki, R. (2015). Pyrolysis of *Mesua ferrea* and *Pongamia glabra* seed cover: Characterization of bio-oil and its sub-fractions. *Bioresource Technology, 178,* 83–89.
13. Prakash, P., & Sheeba, K. N. (2016). Prediction of pyrolysis and gasification characteristics of different biomass from their physico-chemical properties. *Energy Sources, Part A: Recovery, Utilization, and Environmental Effects, 38*(11), 1530–1536.
14. Strezov, V., Moghtaderi, B., & Lucas, J. (2003). Thermal study of decomposition of selected biomass samples. *Journal of Thermal Analysis and Calorimetry, 72,* 1041–1048.

15. Rafiq, M. K., Bachmann, R. T., Rafiq, M. T., Shang, Z., Joseph, S., & Long, R. (2016). Influence of pyrolysis temperature on physico-chemical properties of corn stover (*Zea mays* L.) biochar and feasibility for carbon capture and energy balance. *PLoS ONE, 11*(6), e0156894.
16. Gao, Y., Yang, Y., Qin, Z., & Sun, Y. (2016). Factors affecting the yield of bio-oil from the pyrolysis of coconut shell. *Springer Plus, 5,* 333.
17. Mayakaduwa, S. S., Vithanage, M., Karunarathna, A., Mohan, D., & Ok, Y. S. (2016). Interface interactions between insecticide carbofuran and tea waste biochars produced at different pyrolysis temperatures. *Chemical Speciation & Bioavailability, 28,* 1–4.
18. Gray, M., Johnson, M. G., Dragila, M. I., & Kleber, M. (2014). Water uptake in biochars: The roles of porosity and hydrophobicity. *Biomass and Bioenergy, 61,* 196–205.
19. Sut, D., Chutia, R. S., Bordoloi, N. J., Narzari, R., & Kataki, R. (2016). Complete utilization of non-edible oil seeds of *Cascabela thevetia* through a cascade of approaches for biofuel and by-products. *Bioresource Technology, 213,* 111–120.
20. Liu, Z., Niu, W., Chu, H., Zhou, T., & Niu, Z. (2018). Effect of the carbonization temperature on the properties of biochar produced from the pyrolysis of crop residues. *BioResources, 13*(2), 3429–3446.

Influence of Temperature on Quality and Yield of Pyrolytic Products of Biofuel Process Wastes

Samarjit Gogoi, Rumi Narzari, Neonjyoti Bordoloi, Nilutpal Bhuyan, Debashis Sut, Lina Gogoi and Rupam Kataki

Abstract With the ambitious biofuel project launched by the government of India, a good number of non-edible oil seeds have been explored as a potential candidate to produce biodiesel. However, in this attempt, a massive amount of solid byproducts is also generated (seed cover and de-oiled cake). These byproducts if not utilized add to the wastes and may end up in open burning. Thus complete utilization of biomass resources are vital for maximizing the efficiency, sustainability and economics of biofuel production. In this work, we have investigated the pyrolysis properties of *Kayea assamica* (KA) seedcake (a biofuel process waste) using TGA analysis. The slow pyrolysis reaction was performed with 40 °C/min heating rate. The physico-chemical characteristics of *Kayea assamica* were investigated by bomb calorimetry, elemental analysis (CHN), SEM and FT-IR spectroscopy. The proximate and ultimate analysis was conducted using standard ASTM methodology. The variation of product yield with the pyrolysis temperature was also observed. Energy performance system of the pyrolysis system was also evaluated. BET analyses of biochar were also studied for the preparation of activated carbon.

Keywords Renewable energy · Energy performance · Product yield · Biochar

S. Gogoi · R. Narzari · N. Bordoloi · N. Bhuyan · D. Sut · L. Gogoi · R. Kataki (✉)
Department of Energy, Tezpur University, Tezpur 784028, Assam, India
e-mail: rupamkataki@gmail.com

S. Gogoi
e-mail: samarjitipl123@gmail.com

R. Narzari
e-mail: narzarirumi@gmail.com

N. Bordoloi
e-mail: neon_bordoloi@yahoo.co.in

N. Bhuyan
e-mail: nilutpal6bhuyan@gmail.com

D. Sut
e-mail: debashissut07@gmail.com

L. Gogoi
e-mail: lina.dbr@gmail.com

© Springer Nature Singapore Pte Ltd. 2020   129
S. K. Ghosh (ed.), *Energy Recovery Processes from Wastes*,
https://doi.org/10.1007/978-981-32-9228-4_11

1 Introduction

Environmental pollution caused due to the burning of conventional fuels to meet the exponential growth in energy demand, energy security, energy sustainability, and energy accessibility are some of the global issues currently faced by the humanity, and a shift over to renewable energy resources like biomass, solar, wind, etc. can be a part of the solution in this regard. With the growth rate of 2.5% per annum, renewable and nuclear energy becomes the fastest growing energy resources [1]. The future relies on the growth of renewable sources for the development of a secure and sustainable energy system. Renewable sources account for 13% of the world total primary energy in 2013 [2].

India is bestowed by profuse amount of renewable resources. As stated by the Ministry of New and Renewable Energy (MNRE), the potential of India to produce renewable energy is over 245 GW. It produces 400–500 millions of biomass per year and provides 32% of the total primary energy consumption [3]. Biomass refers to all the organic compounds which are combustible in nature. Among the several available technologies toward deriving energy from biomass, the thermochemical approach has gained tremendous interest [4, 5]. The thermochemical methods can be divided into three class: (i) gasification, (ii) combustion, and (iii) pyrolysis. Out of these, pyrolytic conversion offers twin advantages of possibility of using diverse groups of biomasses as feedstock, and diverse products yields. It breaks down the polymeric substances into solid (bio)char, liquid (bio-oil) and gaseous product with lower molecular weight [6]. Biochar can be applied as a soil amendment, used as solid fuel, and catalyst during chemical manufacturing. The gaseous product is composed of CH_4, CO_2, H_2, N_2 and traces of volatile organic compounds which make it suitable for industrial application as an energy source [7, 8]. The liquid fractions used as fuel for furnaces and boilers. Although there are some demerits of using bio-oil without any modification due to the presence of oxygenated phenolic and ester constituents, water, and some corrosive organic compounds.

Demirbas [9] studied temperature (500–1200 K) effect on pyrolysis products hazelnut, walnut, almond, and sunflower. Their study reveals that char yield was inversely related to a pyrolysis temperature. Soetardji et al. [10] performed pyrolysis on jackfruit peel waste at varying temperature. They found that bio-oil yield was maximum at temperature 550 °C. The GC-MS reports for bio-oil states the presence of ketones, alcohols, furans, acid and esters. Shadangi et al. [11] studied the effect of co-pyrolysis of karanja and niger seed with polystyrene on liquid fuels. They reported that polystyrene enhanced the yield of bio-oil. At 2:1 mixing ratio highest calorific value of oil was reported. Pradhan et al. [12] investigated the pyrolytic behavior of mahua seed. The maximum bio-oil yield was reported as 49% at 525 °C, while the maximum char yield was 18%. The calorific values of the pyrolytic oil and biochar were found to be 39.02 and 26.05 MJ/kg respectively. Sokoto and Bhaskar [13] performed the kinetic study of castor seed de-oiled cake (*Ricinus communis*) by employing the thermogravimetric procedure. The study revealed that the temperature range for maximum decomposition was 329.94–355 °C. The proximate and ultimate analytical result indicated the viability of castor de-oiled seedcake as a feedstock for thermochemical treatment to produce biofuel. The studies above establishes that biowastes including de-oiled seed cakes are suitable feedstocks for their pyrolytic valorization to various products.

Kayea assamica (KA) is a non-edible oil seed bearing tree species widely available in the forest of Assam, India. It is an evergreen tree characterized by light brownish gray bark. Presence of β-farnesene (a biological toxin) in the fruits, makes it unsuitable for cattle feed. Therefore, the KA de-oiled seedcake (KASC) remaining as a biofuel process wastes after utilization of KA oil for biodiesel production, can be suitably used as a feedstock for thermochemical conversion process to various products [14].

The aim of the current study is to investigate the pyrolytic behavior of the non-edible de-oiled seed cakes of *Kayea assamica* (KASC) at different temperature and to investigate the effects of pyrolysis temperature on the distribution of product yield. Characteristic properties including proximate–ultimate analysis, surface structure, functional groups, and Calorific value of the raw biomass, biochar, and bio-oil obtained in a fixed bed pyrolysis reactor were evaluated.

2 Materials and Methods

2.1 Raw Material

KASC was obtained from the Biofuel Lab of the Energy Department of Tezpur University. The seedcake sample was then passed through 36 B.S (0.42 mm) sieve and the resultant sample was oven dried at 105 °C to remove moisture. After moisture removal, the sample is kept in an airtight container.

2.2 Pyrolysis Experiment

Biomass sample weighing 10 g was subjected to pyrolysis in a fixed bed reactor at four different temperatures is. 350, 450, 550 and 650 °C in N_2 atmosphere. The pyrolysis processes were carried out with a heating rate of 40 °C/min. Product yields during pyrolysis were calculated on a dry basis for the biomass samples. The total yield of char, liquid and gas were given by the following equations:

$$\text{Biochar yield (\%)} = (\text{wt. of biochar produced/wt. of feedstock}) \times 100 \quad (1)$$

$$\text{Liquid yield (\%)} = (\text{wt. of liquid portion recovered/wt. of the feedstocks}) \times 100 \quad (2)$$

$$\text{Gas yield } (\%) = 100 - (\text{biochar yield\%} + \text{liquid yield\%}). \tag{3}$$

2.3 Characterization

2.3.1 Characterization of Biomass and Biochar

Proximate analysis of KASC and biochar were performed using ASTM methods (D3173 for moisture, D3174 for ash, D3175 for volatile matters). The fixed carbon content was calculated by difference. Elemental composition and calorific value was evaluated according to ASTM D2015 method using a elemental analyzer (model: Euro EA 3000) and an auto bomb calorimeter (model: 5E-1AC/ML) respectively. pH and electrical conductivity were measured using pH meter (EUTECH pH 700) and a conductivity meter (Digital TDS/Conductivity Meter MK509) as described elsewhere [15]. SEM was performed on a Jeol, JSM-6290LV instrument. FT-IR spectrum of biomass and biochar was performed on a Nicolet IR spectrometer with a spectral range of 4000–400 cm^{-1}.

2.3.2 Characterization of Bio-Oil

The elemental analysis, FT-IR, GCMS and calorific value of bio-oil obtained at highest temperature were determined using methods as described elsewhere [15].

2.4 Thermal Analysis

Thermal behavior of the biomass was investigated by using Perkin Elmer STA 6000 TG analyzer. The sample was heated from 30 to 800 °C at a heating rate of 10 °C/min under a flow of high purity N$_2$ gas (99.99%) at a flow rate of 30 ml/min.

2.5 Biomass Composition

Biomass is composed of cellulose, hemicelluloses, and lignin. Therefore compositional analysis was done to quantify the number of individual constituents. Using TAPPI Standard [16] extractives were removed from the sample and calculated. To determine the hemicellulose content 1 g of the extractive-free sample was added to 150 ml of 500 mol/m^3 NaOH and the mixture was boiled for 3.5 h. After dying the residue was weighed. The hemicelluloses wt% was determined from the difference between the weight of the sample before and after the treatment. TAPPI Standard [17] was employed to determine the lignin content. All the experiments were conducted in triplicate. The cellulose content (%w/w) was calculated by difference [18].

3 Results and Discussion

3.1 Biomass Characteristics

The proximate analysis was widely carried out to evaluate the fuel quality. The outcomes were tabulated below. From the proximate analysis, it was observed that the volatile matter content was very high (72.34%) while the fixed carbon content was 8.97%. Due to easy decomposition of volatile matter into smaller organic compounds, the high volatile content of the feedstock is preferred for pyrolysis. High ash content implies lower energy conversion efficiency [19]. Higher volatile matter content and low ash, and moisture contents made this sample suitable for application in gasifier and pyrolysis.

Elemental composition of KASC reveals that the major constituent in the biomass is oxygen (46.88%), followed by carbon (43.23%), hydrogen (8.23%) and very trace amount of nitrogen (1.66%). This implied that during pyrolysis less amount of nitrous and sulfur oxides would be produced. The H/C and O/C ratios for KASC were 1.623 and 0.821 respectively. Due to the higher amount of oxygen in KASC biomass, it showed lower calorific value (17.82 MJ/Kg) (Table 1).

Table 1 Proximate and ultimate analysis of KASC

Analysis type	KASC
Proximate analysis (wt%)	
Moisture	8.35
Ash	10.34
Volatile matter	72.34
Fixed carbon[a]	8.97
Ultimate analysis (wt%)	
Carbon	43.23
Hydrogen	8.23
Oxygen[a]	46.88
Nitrogen	1.66
H/C molar ratio	1.62
O/C molar ratio	0.821
Empirical formula	$C_{30}H_{69}NO_{25}$
Calorific value (MJ/kg)	17.82

[a]By difference

Table 2 Component analysis of de-oiled seedcake sample (%w/w)

Sample name	Extractives	Cellulose	Hemicelluloses	Lignin
KASC	16.45 ± 0.05	22.56 ± 0.58	20.25 ± 0.37	40.74 ± 0.31

3.2 Component Analysis of Biomass

The cellulose, hemicellulose, lignin and extractives contents percentage were determined from the biochemical and inorganic element analysis. The yields of pyrolysis product were found to be proportional to the cellulose, hemicellulose and lignin content of the sample. Biochar yield is dependent on lignin content of the feedstock; i.e. higher the lignin content, higher is biochar yield [19] (Table 2).

3.3 FT-IR Spectra of Biomass

From KASC FT-IR, the stretching vibration at 3079 cm^{-1} indicated the presence of alkenes. The vibration at 1561 cm^{-1} were the symmetrical and asymmetrical stretching of C=C, prove the presence of aromatic compounds. The small peak at 1124 cm^{-1} showed the presence of C–O functional groups of alcohol, alkyl halide or ether. The vibration at 875 cm^{-1} represented C–H stretching vibration (Fig. 1).

Fig. 1 FT-IR of KASC

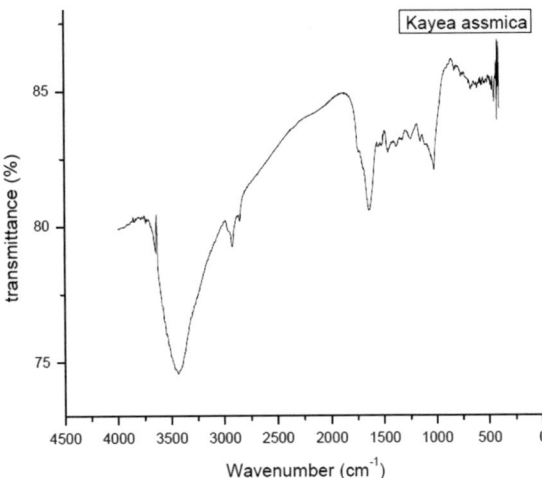

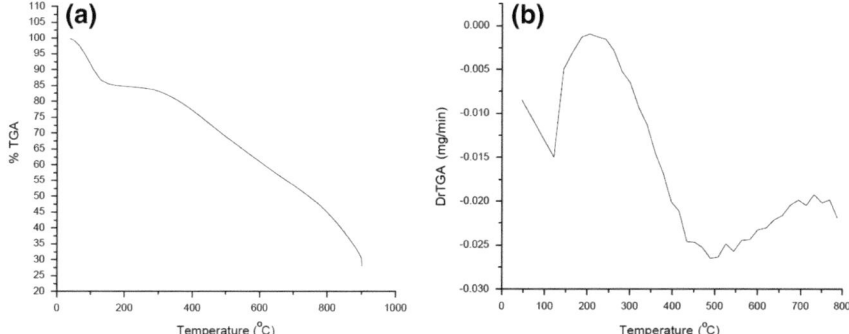

Fig. 2 **a** TGA curve of KASC. **b** DTGA graph of KASC

3.4 Thermal Analysis

Figure 2a, b represents the TG and DTG curves for KASC. The thermogravimetric analysis of the KASC was carried out in N_2 atmosphere with a heating rate of 10 °C/min. From the TG graph biomass pyrolysis can be divided into various regions describing gradual decomposition in each stage. The first stage represents the removal of moisture content and lighter volatile constituents, followed by breaking down of hemicellulose; decomposition of lignin and cellulose and at the end constant decomposition of lignin [20, 21]. The initial weight loss was observed at temperature range 28–110 °C due to the loss of moisture. In this region, a loss of 8.35 wt% were observed for KASC. Loss of volatile matter was marked with a sharp fall of biomass weight at 280 °C. Beyond this temperature the weight loss was slow and gradual, owing to the burning of char which continued to 700 °C. Hemicellulose thermally decomposed at 220–315 °C, followed by cellulose at 315–400 °C and lignin at wide temperature range of 180–900 °C [13, 22]. From the TG curve it was observed that most of the mass was lost at temperature 700 °C, so a temperature range of 500–650 °C would be suitable for bio-oil production though pyrolysis.

3.5 Pyrolysis Products Yield

Pyrolysis of KASC was performed at four temperatures ranging from 350 to 650 °C (100 °C interval) with 40 °C/min heating rate. To evaluate the pyrolytic condition at which maximum bio-oil is produce the effect of temperature on the products yield was investigated. The char yield decrease from 39.45 to 34.47% due to temperature increment from 350 to 650 °C (Fig. 3), owing to the higher breakdown of primary molecules of biomass as temperature increased. Also, both secondary decomposition reactions and removal of volatile matter were responsible for the decrease in char production at higher temperature [23]. The increase in temperature caused secondary

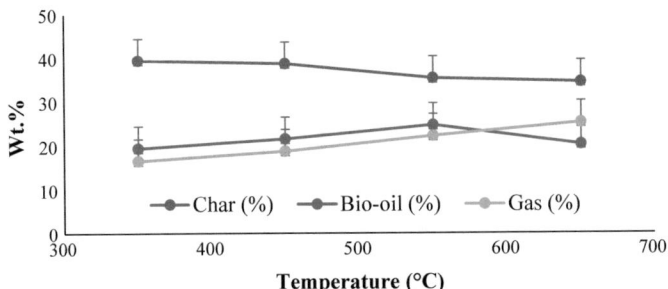

Fig. 3 Product yield % of KASC

cracking of the pyrolytic gases which led to higher bio-oil production at higher temperature [24]. Higher pyrolysis temperature favored the higher yield of bio-oil because it enhanced the volatile availability and reduces time remained for secondary cracking.

It was observed that with the increasing temperature there was a rise in the yield of gaseous product. The gas product yield increased from 16.56 to 27.26% for KASC which can be associated with the secondary cracking of vapors at higher temperatures.

3.6 Characterization of Biochar

The proximate and elemental analytical results for KASC biochars has been summarized in Table 3. The result indicated that the volatile content of biochar decreased from 30.56 to 26.66% for KASC. Biochar exhibited very lower amount of volatile compared to original biomass. The fixed carbon content and ash content also proportionately increased with temperature. The escalation in the cationic concentration of Mg, Ca, K, S, P, and Zn is responsible for the subsequent increase in ash residue [25]. The increased dehydration of the biochar at elevated temperature indicates less hydrophilic biochar surface at a higher temperature. The increase in the calorific value of biochar with a rise in production temperature represents the effect of the increased concentration of carbon.

The elemental composition revealed that the amount of carbon in biochar amplified whereas the oxygen and hydrogen content in biochar declined with rising temperature. The increase in carbon was attributed to increases in the degree of carbonization [26]. The decline in the fraction of O and H content was due to cleavage and breaking of weaker bonds in the biochar structure [27]. The loss of moisture, decarboxylation, and decarbonylation with rise in temperature might be a possible reason for decline in H/C and O/C proportion of biochar [24].

With the rise in final pyrolysis temperature a change in pH and EC of the biochar was observed which may be due to the loss of acid functionality, and separation of alkali salts from organic material [15]. The heterogeneous and amorphous structure

Table 3 Physicochemical properties of biochar from KASC

Properties	Biocharfrom KASC			
Temperature (°C)	350	450	550	650
Moisture content (%)	8.45	7.89	7.56	7.50
Volatile content (%)	30.56	29.67	28.89	26.66
Ash content (%)	5.67	6.65	6.90	7.20
Fixed carbon[a] (%)	55.32	55.79	56.65	58.64
Calorific value (MJ/Kg)	17.56	21.32	22.54	24.67
C (%)	70.25	72.32	75.09	76.12
H (%)	2.85	2.56	2.29	2.22
O[a] (%)	25.33	23.78	21.00	20.18
N (%)	1.57	1.34	1.65	1.48
H/C	0.49	0.42	0.37	0.35
O/C	0.27	0.25	0.21	0.20
pH	9.48	9.56	9.69	10.12
EC (dSm^{-1})	1.79	2.56	3.12	3.89
BET surface area (m^2/g)	–	6.728	12.457	–

[a]By difference

in biochar were also observed from SEM images. The char obtained at a higher temperature had a more irregular and distorted structure compared to the char at low temperature (Fig. 4).

(a)   **(b)**

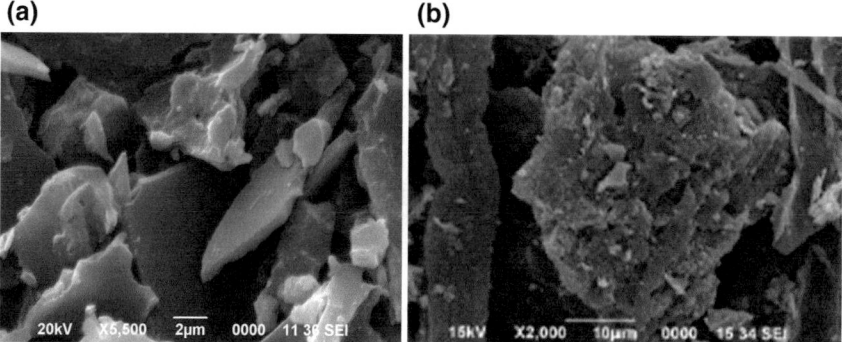

Fig. 4 SEM images of biochar from KASC at **a** 350 °C and **b** 650 °C

3.7 FT-IR Spectra of Biochar

The functional groups for the biochar produce were studied from the FT-IR spectra of KASC (Fig. 5). The peak obtained at 3412 cm^{-1} indicated the presence of phenols, alcohol, water and N–H$_2$. However, the band intensities gradually declined with the rising temperature caused by accelerated dehydration reaction. The presence of aliphatic CH$_3$ stretching vibration is represented by the presence of band at 3079 cm^{-1}. The band at 2360 cm^{-1} showed the presence of cyanides and alkynes. The intensity of C=C stretch at 1561 cm^{-1} attributed to carboxyl and carbonyl group decreased with increasing pyrolysis temperature. The weak bond frequencies around 1050 and 1350 cm^{-1} shows the presence of phenols, alcohols (primary, secondary and tertiary), ethers and esters with C–O stretch and O–H deformation. At 875 cm^{-1} frequency, with =C–H and C–Cl stretching indicated the presence of alkenes and alkyl halide [23].

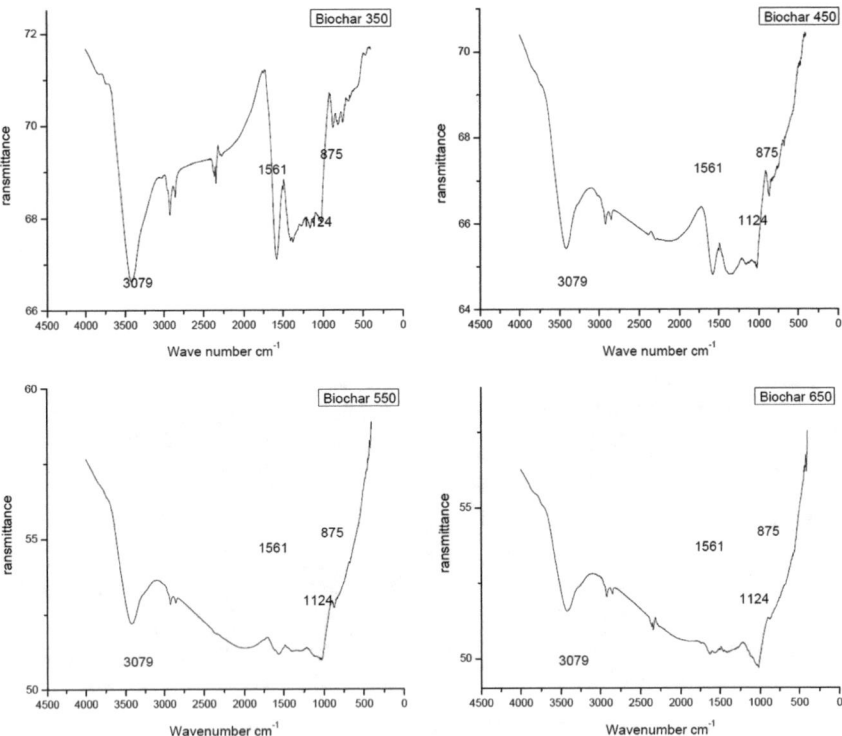

Fig. 5 FT-IR spectra of KASC biochar at four different temperatures

3.8 Characterization of Bio-Oil

Table 4 illustrates the elemental composition and calorific value of the liquid product which helped in assessing and determining the feasibility of bio-oil as liquid fuel. Low oxygen content in the bio-oil (32.23%) compare to its feedstock (46.88%) was revealed from the elemental analysis which makes it suitable transportation fuel. The high calorific value of bio-oil implies better fuel properties.

The O–H, N–H stretching vibration at 3461 cm^{-1} indicated the presence of water, N–H$_2$ group in bio-oil obtained from KASC. The peak obtained at 2964 cm^{-1} attributed to the presence of alkanes. The stretching of C=C vibration at 1624 cm^{-1} indicated the existence of alkenes. The presence of ether and alcohol functional group is confirmed by the C–O stretch and O–H bend at 1124 cm^{-1} (Fig. 6).

Table 4 Ultimate analysis of bio-oil obtained from KASC

Parameters	KASC
C (%)	57.78
H (%)	6.87
N (%)	3.12
O^a (%)	32.23
H/C	1.43
O/C	0.42
Empirical formula	$C_{21.6}H_{30.8}NO_9$
Acid Number	36.77
Calorific value (MJ/Kg)	24.80

^{a}By difference

Fig. 6 FT-IR spectra of KASC bio-oil at 550 °C

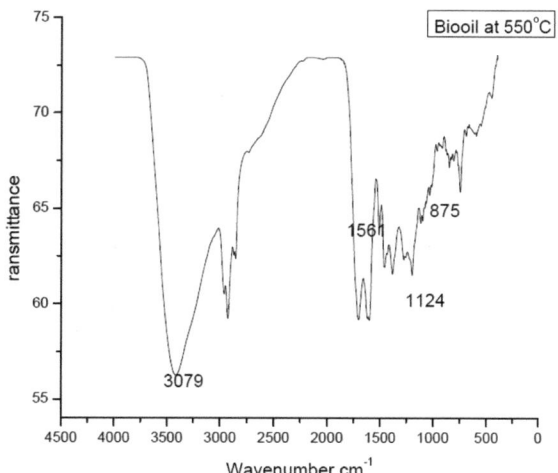

From the comparative study of the data obtained from the bio-oil chromatogram (Fig. 7) with standard (NIST 98 spectrum library) the compounds present in the bio-oil were identified. Due to the intricate composition of bio-oil, it was difficult to perfectly separate all the peaks. Table 5 summarizes the identified compounds with area percentage and retention time (RT) in minutes. According to Wang et al. [28] hemicelluloses and cellulose interaction in biomass, effect the formation of sugars negatively while that of ketones and furans positively. During pyrolytic degradation of biomass, the phenylpropane units of lignin contribute to the production of more quantities of phenols and their derivatives. In the present investigation, most of the phenolic compounds such as phenol, phenol-2,4 dimethyl, phenol-2-methyl, phenol-4-methyl, and phenol-2-methoxy derived from lignin decomposition were present in a greater amount.

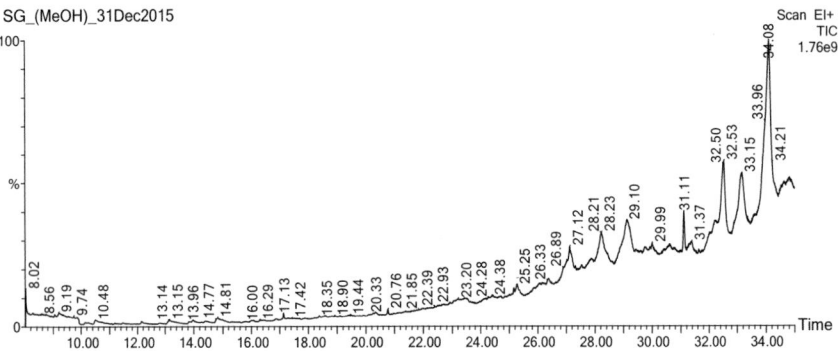

Fig. 7 Gas chromatogram of KASC bio-oil

Table 5 Retention time and compound name present in bio-oil obtained at 550 °C

S. no.	RT (min)	Area	Compound name
1	13.15	0.41	Phenol, 2 methyl
2	14.80	0.72	Phenol, 2 methoxy
3	23.19	0.73	Hydroquinone
4	23.41	1.04	Phenol, 2,4 dimethyl
5	24.71	0.20	2-cyclopenten-1-one,2-methyl-
6	25.15	0.23	Phenol, 4 methyl-
7	26.38	1.04	Benzaldehyde,2 methyl-
8	28.22	8.36	Phenol
9	29.13	12.38	Phenol,4 ethyl
10	30.00	2.79	3-furanmetanol
11	31.38	1.35	Pyridine
12	32.52	7.38	Phenol, 4 methyl

4   Conclusion

The current investigation of pyrolytic valorization of a biofuel process waste i.e. KASC was carried out in a fixed bed reactor at four different temperatures, ranging from 350 to 650 °C with an increment of 100 °C. This study established that the final pyrolysis temperature has a major effect on the yield and physicochemical characteristics of the pyrolysis products. Biochar yield decreased with the increase in pyrolysis temperature. The pH of the biochar also increased with the temperature and basic in nature which is suitable for soil amendment in the northeastern region of India. The yield of bio-oil was maximum at 550 °C with a caloric value of 24.80 MJ/kg. The GC-MS results revealed that the bio-oil is mainly composed of phenolic compounds.

Acknowledgements Tezpur University and Guwahati Biotech Park were acknowledged by the authors for providing facilities to conduct various analysis. R. Narzari, and D. Sut also acknowledge the fellowship grants from UGC.

References

1. IEO report. (2013). International energy outlook. Accessed on April 4, 2018.
2. International energy agency. www.iea.org/aboutus/faqs/renewableenergy/. Accessed on May 8, 2018.
3. Renewable energy in India: status and growth. FY-2013.
4. Yee, L. H. (2010). Initiatives to promote renewable energy. Available at: http://www.biomass-sp.net/press/October/20.pdf.
5. Idris, S. S., Rahman, N. A., & Ismail, K. (2012). Combustion characteristics of malaysian oil palm biomass, sub-bituminous coal and their respective blends via thermogravimetric analysis (TGA). *Bioresource Technology, 123,* 581–591.
6. Ceylan, S., & Topcu, Y. (2014). Pyrolysis kinetics of hazelnut husk using thermogravimetric analysis. *Bioresource Technology, 156,* 182–188.
7. Jahirul, M. I., Rasul, M. G., Chowdhury, A. A., & Ashwath, N. (2012). Biofuels production through biomass pyrolysis-a technological review. *Energies, 5*(12), 4952–5001.
8. Meier, D., Beld, B. V., Bridgwater, A. V., Elliot, D. C., Oasmaa, A., & Preto, F. (2013). State-of-the-art of fast pyrolysis in IEA bioenergy member countries. *Renewable and Sustainable Energy Reviews, 20,* 619–641.
9. Demibas, A. (2006). Effect of temperatures on the pyrolysis product of four nut shells. *Journal of Analytical and Applied Pyrolysis, 76,* 285–289.
10. Soetardji, J. P., Widjaja, C., Djojorahardjo, Y., Edi, F., Soetaredjo, F. E., & Ismadji, S. (2014). Bio-oil from jackfruit peel waste. *Procedia Chemistry, 9,* 158–164.
11. Shandangi, K. P., & Mohanty, K. (2015). Co-pyrolysis of Karanja and Niger seeds with waste polystyrene to produce liquid fuels. *Fuel, 153,* 492–498.
12. Pradhan, D., Singh, R. K., Bendu, H., & Mund, R. (2016). Pyrolysis of Mahua seed-production of bio-fuel and its characterization. *Energy Conversion Management, 108,* 529–538.
13. Sokoto, A. M., & Bhaskar, T. (2018). Pyrolysis of waste castor seed cake: A thermo-kinetics study. *European Journal of Sustainable Development Research, 2*(2), 1–12.
14. Ramachandran, S. K. S., Sudheer, L., Christian, C. R., & Soccol, Pandey A. (2007). Oil cakes and their biotechnological applications—A review. *Bioresource Technology, 98,* 2000–2009.

15. Bordoloi, N., Narzari, R., Chutia, R. S., Bhaskar, T., & Kataki, R. (2015). Pyrolysis of *Mesua-ferrea* and *Pongamiaglabra* seed cover: Characterization of bio-oil and its sub-fractions. *Bioresource Technology, 178,* 83–89.
16. TAPPI Standard T 204 om-88. (1988a). Solvent extraction of wood and pulp. In *TAPPI Test Methods 1994–1995.* TAPPI, Technology Park, Atlanta, GA.
17. TAPPI Standard T 222 om-88. (1988b). Acid-insoluble lignin in wood and pulp. In *TAPPI Test Methods 1994–1995.* TAPPI, Technology Park, Atlanta, GA.
18. Blasi, C. D., Signorelli, G., Russo, C. D., & Rea, G. (1999). Product distribution from pyrolysis of wood and agricultural residues. *Industrial & Engineering Chemistry Research, 38*(6), 2216–2224.
19. Rajamohan, S., & Kasimani, R. (2018). Analytical characterization of products obtained from slow pyrolysis of *Calophylluminophyllum* seed cake: Study on performance and emission characteristics of direct injection diesel engine fuelled with bio-oil blends. *Environmental Science and Pollution Research, 25,* 9523–9538.
20. Yang, H., Yan, R., Chen, H., Lee, D. H., & Zheng, C. (2007). Characteristics of hemicellulose, cellulose and lignin pyrolysis. *Fuel, 86,* 1781–1788.
21. Sanchez-Silva, L., Lopez-Gonzalez, D., Villasenor, J., Sanchez, P., & Valverde, J. L. (2012). Thermogravimetric–mass spectrometric analysis of lignocellulosic and marine biomass pyrolysis. *Bioresource Technology, 109,* 163–172.
22. Barneto, A. G., Carmona, J. A., Alfonso, J. E. M., & Serrano, R. S. (2010). Simulation of the thermogravimetry analysis of three non-wood pulps. *Bioresource Technology, 101,* 3220–3229.
23. Chutia, R. S., Kataki, R., & Bhaskar, T. (2014). Characterization of liquid and solid product from pyrolysis of *Pongamiaglabra* deoiled cake. *Bioresource Technology, 165,* 336–342.
24. Gogoi, D., Bordoloi, N., Goswami, R., Narzari, R., Sakia, R., Sut, D., et al. (2017). Effect of torrefaction on yield and quality of pyrolytic products of arecanut husk: An agro-processing wastes. *Bioresource Technology, 242,* 36–44.
25. Melo, L. C. A., Coscione, A. R., Abreu, C. A., Puga, A. P., & Camargo, O. A. (2013). Influence of pyrolysis temperature on cadmium and zinc sorption capacity of sugar cane straw-derived biochar. *BioResources, 8*(4), 4992–5004.
26. Chen, Y. Q., Yang, H. P., Wang, X. H., Zhang, S. H., & Chen, H. P. (2012). Biomass-based pyrolytic polygeneration system on cotton stalk pyrolysis: Influence of temperature. *Bioresource Technology, 107,* 411–418.
27. Demirbas, A. (2004). Effects of temperature and particle size on bio-char yield from pyrolysis of agricultural residues. *Journal of Analytical and Applied Pyrolysis, 72*(2), 243–248.
28. Wang, S., Guo, X., Wang, K., & Luo, Z. (2011). Influence of the interaction of components on the pyrolysis behavior of biomass. *Journal of Analytical and Applied Pyrolysis, 91,* 183–189.

Recycling Industrial Waste for Production of Bioethanol

Swagata Das, Shubhalakshmi Sengupta, Papita Das and Siddhartha Datta

Abstract In need to overcome the environmental impacts and dearth of fuels, eco-friendly technologies such as biofuels are being developed. Researchers are putting in effort to convert huge amount of lignocellulosic wastes to biofuels such as bioethanol. The main aim for all the second generation biofuels are implying waste management and developing eco-friendly products using the wastes. These wastes are the ways to sustainable waste management. The proper handling of these wastes is too essential, as these cause a threat to the environment. Hence, experimental researches have been focussed on the production of cellulosic bioethanol. Jute caddies were taken up as a potential source for utilizing it for the production of bioethanol. Jute contains a large amount of cellulose, about 60%, which is the major constituent for the conversion to bioethanol. Cellulose is a hard crystalline structure and therefore is subjected to different chemical pretreatments for its degradation and release of fermentable sugars such as glucose, xylose and arabinose as the predominant sugars. Pretreatments such as alkali and alkali plus dilute acid treatment were done followed by enzymatic saccharification using *Aspergillus niger*. Fermentation was the next crucial step for the consumption of these fermentable sugars (simple and complex) by microorganism, for production of bioethanol. *Saccharomyces cerevisiae* (yeast) was used to study the effect on the uptake of these sugars and conversion to bioethanol. The yield was 27% ethanol in this experiment.

S. Das (✉) · S. Sengupta · P. Das · S. Datta
Department of Chemical Engineering, Jadavpur University, Kolkata, India
e-mail: dasswagata600@gmail.com

S. Sengupta
e-mail: sengupta.shubha@gmail.com

P. Das
e-mail: papitasaha@gmail.com

S. Datta
e-mail: sdatta_che@rediffmail.com

© Springer Nature Singapore Pte Ltd. 2020
S. K. Ghosh (ed.), *Energy Recovery Processes from Wastes*,
https://doi.org/10.1007/978-981-32-9228-4_12

1   Introduction

Lignocellulosic biomass is rich in cellulose, which on degradation by chemical or biological treatment releases fermentable sugars. Bioethanol is produced from either starch containing substances or lignocellulosic biomass. The use of bioethanol was first put in practice by the IC engines [1]. Biofuels have been utilized as transport fuels since the commencement of the IC engine. Bioethanol has often been termed as automobile fuel. With the exhaustible fuels crisis, bioethanol has become a major alternative as a substitute for other exhaustible fuels. The research sectors are developing different economically feasible and profitable ways for production of bioethanol. "2-G biofuels" have taken over 1-G biofuels as they are focussed on converting waste to useful environment friendly products [2]. In the coming years, biofuels are estimated to take over the market by their cheap price and pollution free nature. The effect of the concerns related to waste management has attracted the concern for 2-G biofuels from lignocellulosic feedstocks [3]. Therefore, the urge to convert industrial waste feedstocks, rich in cellulose content, is the major focus for bioethanol production [4]. The objective of this research findings was to utilize the wastes generated in the best possible way and convert it into useful products which is both cheap and eco-friendly. So, efforts were put in utilizing the huge amount of jute caddies wastes generated every day and converting it into biofuel—bioethanol. Chemical pretreatments were done to loosen the hard structure of cellulose followed by enzymatic saccharification and fermentation.

2   Literature Survey

As bioethanol can be derived from lignocellulosic biomass and wastes from biomass of crop biennial, it enhances the chances of the monetary benefit of farmers and waste management in industries. Bioethanol is widely used as a crucial biofuel for running automobiles worldwide. Because of the limited sources of the non-renewable fuels such as petroleum, biofuels has taken over extensively [5]. Ethanol marks as a closed carbon dioxide cycle as after burning of ethanol, the emitted carbon dioxide is recycled into plant biomass as plants use it to incorporate cellulose during photosynthesis. Bioethanol manufacturing industries uses energy from inexhaustible sources or wastes sources, thereby reducing the net carbon dioxide emission to the air. Ethanol contains about 35% oxygen that provides to total combustion and thereby decreasing particulate emission which many a times poses health hazards [6]. The hazardous nature is decreased by reducing the GHG. Thus, the adaptation of even 10% bioethanol reduces GHG emissions by 15–19%.

3 Materials and Methodology

3.1 Materials

The industrial jute caddies wastes were collected from Mahadeo Jute Mill, Howrah, India. The wastes were manually trimmed for easy handling during experiment.

The chemicals sodium hydroxide, sulphuric acid was purchased from Loba Chemie, India and potassium dichromate and di-nitro salicylic acid were procured from Merck, India.

3.2 Chemical Pretreatments

The following chemical pretreatments were carried out on the non-treated jute caddies waste.

3.2.1 Alkali Pretreatment

The jute caddies waste (5 g) was subjected to alkali treatment to increase cellulose digestibility.

The manually trimmed jute caddies wastes were subjected to 5% (w/v) alkali treatment [7]. The alkali treated jute caddies waste were kept for 20 min and was then washed with water to remove the excess alkali and maintain a pH of 7.00 followed by oven drying for overnight.

3.2.2 Alkali and Acid Pretreatment

The modification to the above pretreatment was followed by treating the alkali treated jute caddies waste with dilute sulphuric acid treatment. Similar process was maintained. The treated sample was kept for 20 min followed by washing away the excess alkali and acid present in it and maintaining a pH of 7.00. It was then oven dried.

3.3 Cellulose Estimation

The cellulose content of the treated as well as the non-treated jute was estimated using the anthrone method [8].

3.4   Enzymatic Saccharification

The microorganism *Aspergillus* sp. was used for the biodegradation of the cellulose structure [9]. The non-treated as well as the treated jute caddies were then incubated at 30 °C for five days. The reducing sugar (glucose) was then estimated using the DNS method [10].

3.5   Fermentation

The universal yeast, which is anaerobic in nature, *Saccharomyces cerevisiae*, was used for fermentation. The yeast caused the fermentation within 3 days at 30 °C and it was then estimated for ethanol yield [11].

3.6   Product Recovery and Estimation

The fermented culture was extracted as ethanol-water content using the Rotary evaporator and was then determined for the amount of alcohol present in the sample using the dichromate test and the reading were generated using UV-Vis spectroscope [12].

4   Results

Non-treated jute caddies

The non-treated jute caddies wastes were subjected to anthrone test for cellulose (Fig. 1) estimation followed by enzymatic saccharification by *Aspergillus* sp. to estimate the amount of glucose released (DNS method) which was the most essential constituent for fermentation. *Saccharomyces cerevisiae* was used in this experiment for fermentation, thereby yielding 11.16% ethanol (Fig. 2).

Alkali treated jute caddies

The alkali treatment which was aimed for increasing the cellulose digestibility (Fig. 1) for better production of ethanol, gave a positive result on the anthrone test used for cellulose estimation. The amount of hexose sugar released was also more than the non-treated jute caddies waste as estimated using the DNS test. In this case, the fermentation caused by the organism was much more and the production of ethanol estimated was 15.65% using the dichromate method (Fig. 2).

Alkali and acid treated jute caddies

The addition of dilute acid to the previous pretreatment for degradation of the cellulose structure and thereby more release of fermentable sugars were found to be

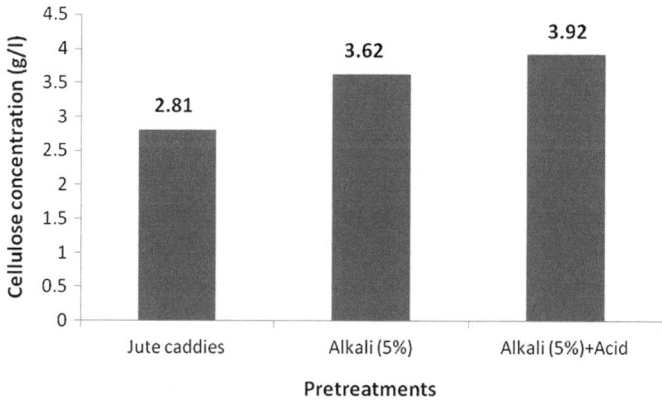

Fig. 1 Cellulose concentration of treated and non-treated jute caddies

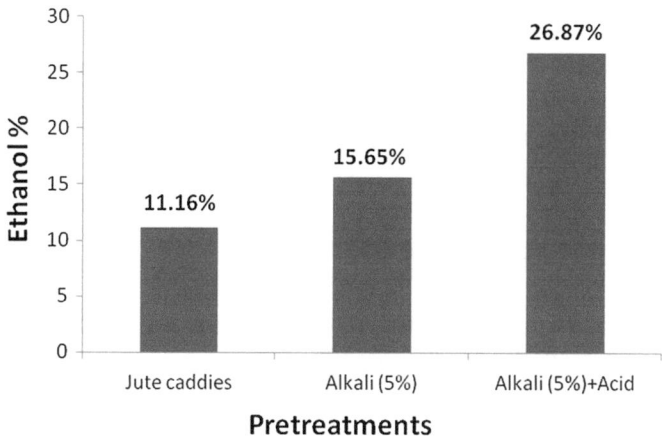

Fig. 2 Ethanol production from non-treated and treated jute caddies

beneficial (Fig. 1). The cellulose content was found to be more than the above two pretreatments. After fermentation the resulting product was 26.87% ethanol (Fig. 2).

4.1 Analysis and Discussions

Non- treated jute caddies

From the anthrone method the amount of cellulose was estimated (Fig. 1). But as the structure of the cellulose is crystalline in nature, it needs chemical or biological pretreatments for its degradation for release of fermentable sugars. As this was non-treated jute caddies waste and was only subjected to enzymatic saccharification

(*Aspergillus* sp.), the structure of cellulose could not be broken so easily. A very limited amount of hexose sugar was released (DNS method) for uptake during fermentation by *Saccharomyces cerevisiae*, thereby resulting only in 11.16% ethanol (Fig. 2).

Alkali

The 5% alkali pretreatment proved to increase the cellulose digestibility using the anthrone method [13] (Fig. 1). This in turn was more effective for the release of the hexose sugar after enzymatic saccharification by *Aspergillus* sp. as reported after the DNS test [14]. The yeast was able to ferment better in case of this treated jute caddies. The ethanol yield was 15.65% (Fig. 2).

Alkali and acid

The modification made to the previous treatment by implying dilute acid treatment was helpful. The dilute sulphuric acid treatment was found to be beneficial in most cases [15]. As seen in the above treatment, alkali treatment increases cellulose digestibility, addition of acid loosens the structure of cellulose more, releasing more of cellulose (Fig. 1) hence more amount of glucose. The glucose was estimated using the DNS method followed by fermentation. The fermented jute caddies waste resulted in 26.87% ethanol as determined by the dichromate test (Fig. 2).

5   Conclusion and Recommendation

The research focused on utilization of jute caddies wastes to bioethanol. Thus to sum up the overall research can be concluded as:

- The two different chemical pretreatments followed by the biodegradation of cellulose, proved to be useful for the production of bioethanol as compared to the non-treated jute.
- The jute caddies wastes which was utilized for this research has been found to be beneficial. It is been reported to produce nearly 27% alcohol on pretreatment. *Saccharomyces cerevisiae* has given a better result on all the cases thereby validating the uptake of the hexose sugars.
- Since this research has focused on waste management and production of eco-friendly products such as biofuels, the used substrate can be reused again for production of bioethanol, thereby completely depleting the source for no further utilization.

Acknowledgements This research was supported by Department of Chemical Engineering, Jadavpur University, by lending their major support by providing necessary instruments required doing this research.

References

1. Wyman, C. E., Cai, C. M., & Kumar, R. (2017). Bioethanol from lignocellulosic biomass. *Encyclopedia of Sustainability Science and Technology*, pp. 1–27.
2. Saini, J. K., Saini, R., & Tewari, L. (2015). Lignocellulosic agriculture wastes as biomass feedstocks for second-generation bioethanol production: Concepts and recent developments. *3 Biotech, 5*(4), 337–353.
3. Tokic, M., Hadadi, N., Ataman, M., Miskovic, L., Neves, D., Ebert, B., Blank, L., & Hatzi-manikatis, V. (2016). Discovery and evaluation of novel pathways for production of the second generation of biofuels (No. POST_TALK).
4. Juneius, C. E. R., & Kavitha, J. (2017). Bioconversion of cellulosic waste into bioethanol—A synergistic interaction of trichoderma viride and *Saccharomyces cerevisiae*. In *Bioremediation and sustainable technologies for cleaner environment* (pp. 201–211). Cham: Springer.
5. Farrell, A. E., Plevin, R. J., Turner, B. T., Jones, A. D., O'hare, M., & Kammen, D. M. (2006). Ethanol can contribute to energy and environmental goals. *Science, 311*(5760), 506–508.
6. Bergeron, P. (2018). Environmental impacts of bioethanol. In *Handbookon bioethanol* (pp. 89–103). Routledge.
7. Manna, S., Saha, P., Chowdhury, S., Thomas, S., & Sharma, V. (2017). Alkali treatment to improve physical, mechanical and chemical properties of lignocellulosic natural fibers for use in various applications. *Lignocellulosic Biomass Production and Industrial Applications*, pp. 47–63.
8. Ribeiro, I. A., Bronze, M. R., Castro, M. F., & Ribeiro, M. H. (2016). Selective recovery of acidic and *Lactonic sophorolipids* from culture broths towards the improvement of their therapeutic potential. *Bioprocess and Biosystems Engineering, 39*(12), 1825–1837.
9. Pierce, B. C., Agger, J. W., Wichmann, J., & Meyer, A. S. (2017). Oxidative cleavage and hydrolytic boosting of cellulose in soybean spent flakes by *Trichoderma reesei* Cel61A lytic polysaccharide monooxygenase. *Enzyme and Microbial Technology, 98,* 58–66.
10. Fu, C. C., Hung, T. C., Chen, J. Y., Su, C. H., & Wu, W. T. (2010). Hydrolysis of microalgae cell walls for production of reducing sugar and lipid extraction. *Bioresource Technology, 101*(22), 8750–8754.
11. Paschos, T., Xiros, C., & Christakopoulos, P. (2015). Simultaneous saccharification and fermentation by co-cultures of Fusarium oxysporum and Saccharomyces cerevisiae enhances ethanol production from liquefied wheat straw at high solid content. *Industrial Crops and Products, 76,* 793–802.
12. Miah, R., Siddiqa, A., Tuli, J. F., Barman, N. K., Dey, S. K., Adnan, N., et al. (2017). Inexpensive procedure for measurement of ethanol: Application to bioethanol production process. *Advances in Microbiology, 7*(11), 743.
13. Darmanto, S., Rochardjo, H. S., Jamasri, & Widyorini, R. (January, 2017). Effects of alkali and steaming on mechanical properties of snake fruit (Salacca) fiber. In *AIP Conference Proceedings* (Vol. 1788, No. 1, p. 030060). AIP Publishing.
14. Coughlan, M. P. (1991). Mechanisms of cellulose degradation by fungi and bacteria. *Animal Feed Science and Technology, 32*(1–3), 77–100.
15. Yoon, S. Y., Han, S. H., & Shin, S. J. (2014). The effect of hemicelluloses and lignin on acid hydrolysis of cellulose. *Energy, 77,* 19–24.

Municipal Solid Wastes—A Promising Sustainable Source of Energy: A Review on Different Waste-to-Energy Conversion Technologies

C. K. Parashar, P. Das, S. Samanta, A. Ganguly and P. K. Chatterjee

Abstract Utilization of Municipal Solid Waste (MSW) as a free resource of energy has gained popularity to reduce the use of conventional fuel. India, as a developing country having a major population residing in rural areas with unplanned societies, unscientific waste management is steeping at enormous rate. This paper highlights several waste-to-energy (WTE) conversion technologies incorporated for the utilization of MSW to produce useful energy. With high end research and development, pyrolysis and gasification processes incorporating enormous temperature and heating rates, has paved the way for the utilization of high density polymeric wastes in generating oil and synthetic gas as products for utilization as fuels having high calorific values. Utilization of kitchen, agricultural and other organic wastes is a promising energy resource to mitigate the dependency on conventional fuels, if scientifically adapted all over the geographical area. In particular, since animal rejects, food waste and agricultural remains cannot be altered, their utilization as fuel is a promising free source of energy. These wastes are fed for biogas production, along with production of alternative manure for farming through bio-methanation and composting. In view of applications in automobiles, aircrafts, and domestic purposes, conversion of methane enriched fuel into bio-diesel paves the way towards sustainable future. As the supply of conventional fuels are facing threat in view of scarcity and pollution caused by their usage, energy from MSW will eradicate the problems of landfills, pollution and other waste management issues, also economically benefiting the society with their livelihood.

C. K. Parashar · P. Das · S. Samanta · A. Ganguly (✉) · P. K. Chatterjee
CSIR Central Mechanical Engineering Research Institute, Durgapur 713209, India
e-mail: amitganguly022@gmail.com

C. K. Parashar
e-mail: chintakparashar@gmail.com

P. Das
e-mail: parthadas.besus@gmail.com

S. Samanta
e-mail: samantasubho@gmail.com

P. K. Chatterjee
e-mail: pradipcmeri@gmail.com

© Springer Nature Singapore Pte Ltd. 2020
S. K. Ghosh (ed.), *Energy Recovery Processes from Wastes*,
https://doi.org/10.1007/978-981-32-9228-4_13

Keywords Municipal solid waste · India · Waste to energy · Sustainable · Society

1 Introduction

The last few decades have seen a massive increase in the amount of municipal solid waste (MSW) in the urban and semi-urban areas of developing nations. Citing an instance, waste generation scenario in Japan is around 1.1 kg/day per inhabitant and about 5.2×10^7 tons/year. In addition to this, the total quantity of industrial waste is estimated around 4×10^8 tons/year [1]. Thus cities are facing a challenge with management of the waste materials due to less effective waste handling methods. This is even growing serious as the composition of the MSW varies from place to place and from time to time. MSW has become a challenge for Urban Local Bodies (ULBs) throughout the world. Sustainable solid waste management (SWM) with financial sustainability is a major challenge in cities of developing countries. Depending upon the country spending capacity; every nation generates a different quantity of waste per capita of MSW. Third world countries generate 250 g per capita per day, whereas developed nations generate up to 2 kg per capita per day. Population explosion coupled with lifestyle improvement and consumption patterns is resulting in significant increase in waste generation. As per one estimate by CPCB (Central Pollution Control Board, New Delhi), in 2010, India generates 60 Million MT of MSW per annum. The growth rate of MSW generation for developing countries is over 5% on an annual basis.

In this regard, quantification of waste is one of the major parameters to be concerned for employment of efficient waste-to-energy conversion system. But problem persists in relation to the quality of waste which varies from place to place and season to season. Also, improper waste handling is an issue as it leads to assorted feed inputs to the WTE conversion systems. Such problem overlooks the recyclable components of the waste which can be easily brought back into the market. Stable segregation system of MSW into different recyclable and non-recyclable components will help in further processing of solid waste. The recyclable material such as metallic parts can be reused easily. The non-recyclable material (i.e. plastic, cotton, rubber, etc.) can be disposed through gasification process to generate energy. A proper segregation system shall help in effective conversion of waste into wealth.

Over the years, several techniques have been developed which utilizes solid waste as an energy resource which also points it out to be a challenging task. The problem lies with the unpredictable composition of waste [2]. According to [3], the element composition of MSW is C: (17–30%), H_2: (1.5–3.4%), O_2: (8–23%), H_2O: (24–34%), ashes: (18–43%) by weight and the average specific combustion heat of MSW is in the range from 5 to 10 MJ/kg. Prior study by [4] indicates gasification (at high temperature) of MSW (as en masse feed input) beaks it down into simpler molecules. The products (along with the by-products) obtained are found to be of high utility.

According to the annual report published during the survey period of 2014–2015, there are 7935 towns in India among which 6166 of them constitute the urban frame of our country, India. Nearly in all states disposal of waste is done by open dumping. 95 landfills have been constructed till 2015 and are spread all over the country but only 14 states have reported monitoring the qualities of the environment like air, water or groundwater at the landfill sites. Being easy to operate, composting process is the preferable option in most of the small towns rather than incineration, gasification or pyrolysis. Industrialized and metro cities have no other options except WTE option which includes incineration and gasification of the waste due to lack of space and energy crisis. A few states like Andhra Pradesh, Gujarat, Goa, Himachal Pradesh, Maharashtra and West Bengal have taken initiative in setting up compost plants. 91 waste-to-energy plants have been reported as established till the reporting year 2014–2015. 648 biogas plants, 3 power plants and 12 RDF projects have been reported in the states of Chandigarh, Delhi, Gujarat, Haryana, Karnataka, Kerala, Maharashtra and Telangana. Efforts are being made for creating awareness among the citizens about the proper disposal and management of solid waste that includes collection, segregation, storage and transportation of the waste [5, 6]. Waste management in developed nations is governed by legislation and looking at the basic approach of legislative tools, they fall into two categories namely "end-of-pipe" regulations and strategic targets.

- End-of-pipe regulations

These are technical regulations and relate to the individual processes in waste treatment and disposal. Emission controls for incinerators are a prime example. Such regulations may be set at national or international standards. These regulations are important to ensure safe operation of waste disposal processes.

- Strategic targets

It defines the way in which solid waste will be dealt with in the future. This legislation has several common threads: it builds on the "hierarchy of solid waste management", and within this it sets targets for recovery and recycling of materials. Much current legislation is also directed at specific parts, rather than the whole, of the municipal waste stream.

1.1 National and International Status of MSW

In India, disposal of Municipal Solid Waste (MSW) poses greater challenges as the projected population as 800,186 in the year 2020, it is estimated that 16,000 tons of MSW would be generated in the year 2020. Nema and Ganeshprasad [7] compared Plasma technologies with conventional waste treatment in an experiment with simulated hospital wastes, at the Institute for Plasma Research in India, using a 50 kW DC transferred arc plasma reactor and results indicted energy recovery is possible with complete neutralization off hazardous element. Dave and Joshi [8] reviewed number of research articles and showed the potential for development of thermal plasma pyrolysis and gasification technologies for waste management and recovery of energy from it. Chatterjee et al. [9] developed a plasma arc pyrolyser for treatment of plastic waste as well as energy recovery from it. Ojha et al. [10] compared various methods of waste management and suggested plasma arc gasification as the most effective way for solid waste management in the developing country. Baidya et al. showed plasma arc gasification as the most sustainable waste to energy process for a country like India using combined AHP and QFD decision-making tools. Baidya et al. proposed a model setup in a composting plant as a case incorporating plasma gasification within the composting plant process line. The study revealed the effectiveness of the proposed model and also showed the sustainability of the process for energy recovery. Pondicherry Urban Development Agency (PUDA) in its detailed report for municipal solid waste disposal through incineration process on Nov-2014 [11]; Eco Save Systems Pvt. Ltd in their report on Municipal Solid Waste Management at Visakhapatnam in Sept-2015 [12]; Tide Technocrats Private Limited in their report on Municipal solid waste collection & transport plan for Gulbarga on Mar-2014 [11] reported major problem for handling municipal solid waste is poor segregation technique. But significant development on segregation of MSW has not yet been carried out explicitly.

Some literature talks about the regulations incorporated by some waste management agencies in US and Europe, and their statistical effects on minimizing waste related problems [13–18]. The studies reveals that formation of management boards in respect to handling the landfills alone, does not solves the MSW generation problem. Proper landfills management reduces the waste by diverting it to some other places. Hence proper recycling regulations depending upon the composition of waste generated/collected needs to be involved. Special emphasis is also given on standardizing the collection and comparison of data related to waste disposal. For instance [18] reported that waste from construction debris should not be compared as waste generation per capita along with the MSW into statistical comparison. The scenario for WTE conversion of MSW and construction debris is different. There are 102 facilities in USA which burns 14% of the total waste generated by 37 million people from 37 of the 50 states in generating 2800 megawatts which is enough for 2.5 million houses. The value of that energy is in excess of $850 million and is produced in facilities that represent a total capital investment of over $10 billion. In 1960s, Japan and many European countries started the massive projects for WTE programme. Transfer of technology to the US first began in late 1960s and early 1970s. However, most projects were not very efficient and were problematic. During mid to late 1970s WTE projects started developing in Saugus, Massachusetts; Pinellas County, Florida; and Ames, Iowa, the areas which were facing the maximum landfill problems [19].

2 MSW Disposal

Employing recycling systems for solid wastes components such as paper, glass, plastic, and so on, not only reduces burden for waste disposal but also utilizes waste as resources which contributes towards sustainable environment. Recyclable waste paper which is the most common ingredient of municipal solid wastes is considered as an "urban ore". Waste papers are considered as the solid recovered fuel (SRF) which is largely contained in municipal solid waste (MSW). Mechanized segregation systems can treat large volume of MSW efficiently with reduced human intervention. But mechanized systems require high capital investment owing to involvement of installation of specific state of the art machineries. In respect to sorting metallic components a lot of literature is available for sorting out metallic components using magnetic action from MSW. Most commonly drum and overhead magnet systems are employed in sorting plants. Their purpose is to remove materials most commonly aluminium, copper, nickel and other metals. Although there are limitations with such systems of sorting out materials only bigger than 10 mm in size but actually it makes the overall MSW segregation simpler by reducing the burden of sorting out bigger metal pieces.

Wastes from the packaging industry are drawing significance attention in terms of quantity of waste generated. As the industry plays a major role in transportation, manufacturing and storage, and almost in all other domains, alternatives are the need of the hour for reducing the environmental effects from packaging wastes. The different types of packaging materials are broadly classified as paper, plastic, cardboard, aluminium and glass.

Waste disposal is generally carried out either by composting (namely aerobic and vermi-composting) or by WTE conversion systems. Employment of WTE units are challenging as the composition of waste is highly diverse. Although in some developed countries these projects have proved to be a success but are still in development stage in many places mainly because of large financial funds required for setting up of these plants, apart from the waste composition diversification factor which gives rise to sustainability issues of the plant. Different MSW treatment methods have been discussed in the subsequent sections [20–22].

2.1 Landfilling

In most of the metropolitan cities, open dumping of MSW is practiced as disposal technique. Such practice gives rise to other environmental problems like, soil contamination, air and water contamination in nearby localities making the zone into an inhabitable place. Since most of the sites are low lying areas, this makes it suitable for water to get trapped with waste and percolates inside it. One of the practices generally observed is levelling of waste after dumping. As per the literature the cost of landfilling is around Rs. 500 to Rs. 1500 per ton of waste [23]. Szabó et al. [24] performed a study for maximizing the benefits of these landfills incorporating solar panels on closed sites. It could be a possible solution for converting the discarded waste into a renewable source of energy by developing solar landfills [25].

Mønster et al. assessed methane emissions from 15 Danish landfills using mobile tracer dispersion method with either Fourier transform infrared spectroscopy, using nitrous oxide as a tracer gas, or cavity ring-down spectroscopy, using acetylene as tracer gas [26].

2.2 Composting

Composting is basically a process in which upon the action of bacteria present in organic matter (or organic waste) converts the waste into high agricultural utility compost, under the influence of hot and moist conditions. This forms the collected waste which was discarded as waste rather into an agricultural fertilizer. Although it demands a framework for implementation and maintenance of the compost formed, but relative to disposal techniques such as incineration it is of high order in terms of productivity. Compost has basically bad odour and proper care is required to prevent diseases approaching out publicly.

Vázquez et al. [27] evaluated the efficiency of home composting programmes and the quality of the compost produced in eight rural areas (880 composting bins). The result obtained was that on average the efficiency in 77% for a compost rate of 126 kg/person per year and the quality was quite good. Home composting helps in production of high quality fertilizers of excellent qualities that can be used in garden or fields.

2.3 Open Sea Dumping

All around the world the oceans are being used as dumping areas for all kinds of wastes including chemical, industrial, radioactive and sewage wastes since decades. This had not been much of an issue since the recent times as it has been well proved that it may lead to extinction of species.

Yang et al. [28] proposed a system of evaluation indexes and methods for evaluating the quality of ocean water in the dumping areas. Commonly used system and index system was taken into consideration for the analysis of the quality of water.

Simonini et al. [29] conducted an experiment to show the long-term dumping effects of dredged material on macro-zoobenthos at 4 different disposal sites in Emilia-Romagna coast. Sediment from harbours was dredged and dumped at the 4 specific locations. The effect was investigated before and after 6 months, 8 months, 2 years and 4 years. The sediments were disposed gradually and no traces were found. The granulometry and %TOC in the sediment remained uninfluenced. Zheng et al. [30] showed the biological effects on mollusks caused by the dumping of waste. The mesocosm technique was used for the same. It was found that high suspended soil content is harmful for mollusks.

2.4 Burning

Burning refers to the burning of waste in open pits or outdoor furnaces. This is also being practiced since decades. Burning releases toxic gases and may produce harmful substances. The smoke may cause health risks to those who are exposed.

Wang et al. [31] studied the atmospheric emissions of heavy metals produced from open burning of MSW in China. MSW contains toxic compounds and when burned in populated areas cause direct exposure to harmful compounds. The further study shows the estimated amounts of typically heavy metals like mercury, arsenic, lead, cadmium, chromium, selenium, copper, zinc and nickel.

2.5 Incineration

Incineration is the process of control and complete combustion, for burning down solid wastes. It helps in energy recovery and destruction of toxic wastes, for instance, waste from hospitals. The temperature in the incinerators varies from 980 to 2000 °C. Incineration process results in the reduction of the original volume of the solid waste by 80–90% the usual by-products of incineration are ash, gaseous and particulate emissions and finally heat energy. The bottom ash formed, is biologically clean and stable which is often used in road building and the construction industry. The flue gas and the fly ash produced in the incineration process are hazardous and are treated by air pollution control equipment. Incineration results in the formation of carcinogenic gases of higher hydrocarbons which are not easy to breakdown.

Abaecherli et al. [32] optimized energy use through systematic short-term management of industrial waste incineration and aims to systematic optimize detailed short-term scheduling for industrial waste incineration to support daily decision-making process. It aims in reducing the utilization of auxiliary fuel which is used to overcome waste related energy incineration process. The study shows reduction potential up to 90% thus reducing CO_2 emission up to 13%. Nabavi-Pelesaraei et al. [33] evaluated the consumption of energy and environmental impacts of incineration and landfill scenarios in Iran. Most energy consumption was found to be related with transportation of waste. Zhou et al. [34] worked on stabilizing the chromates present in the by-products of incineration, i.e. fly ash of formed by the incineration. Ascorbic acid, $NaAlO_2$ and trisodium salt non-anhydrate are the Cr stabilizers and stabilize by reducing Cr (VI) to Cr (III). Hwang et al. [35] estimated the emission of greenhouse gases caused by incineration of waste in Korea. The estimation was done by using US EPA method. The estimation was done across 9 different facilities in Korea each having different operating systems and different NO_x removal systems. Moving grate, stoker, rotary kiln, fluidized bed and kiln and stoker are the different operating systems present. Emission factors of CO_2, CH_4 and N_2O were estimated. Gradus et al. [36] did an effectiveness of cost analysis of plastic waste recycling and WTE incineration process. The plus point of recycling is that is does not have the problem of CO_2 emission. Though the initial cost in incineration is more, it has been found that incineration is more cost-effective. Song et al. [37] analysed the different political, economic, social, technological, environmental and legal aspects in the development incineration. Here MSW treatment status in China has been analysed along with the other aspects in details.

2.6 Pyrolysis

Pyrolysis is a thermochemical decomposition of organic material at high temperatures in the absence of oxygen or any halogen. Pyrolysis generally involves simultaneous changes in chemical composition and physical state which is irreversible. Pyrolysis is generally used in chemical industry. The conversion of wood to charcoal is an example of pyrolysis.

Ansah et al. [38] studied the thermo gravimetric and calorimetric characteristics during co-pyrolysis of MSW. Pyrolysis of PET occurs at higher temperature than components derived from biomass. The plastic pyrolysed at a maximum weight loss rate of 18.5 wt%/min at 420 °C, paper and textile at 10.8 wt% min at 340 °C and wood at 9.9 wt% min at 360 °C during pyrolysis wood required the least amount of heat. Gunasee et al. [39] developed a thermo gravimetric methodology to evaluate the synergistic effects during pyrolysis and combustion of MSW. Here the samples have been heated at 20 °C/min from 25 to 1000 °C under dynamic conditions with the continuous record of their main evolved fragments. By comparing the experimental and calculated weight losses and relative areas of MS peaks synergistic effects were evaluated. Wu et al. [40] investigated oxygen less pyrolysis technique for recycling MSW. It employed Shenwu rotating bed reactor having regenerative radiant tube which provides high temperature. For the same, heat and mass transfer was established by a 3D transient mathematical model that was characterized with simultaneous fluid flow.

2.7 Gasification

Fuel gas comprising of methane, carbon monoxide, carbon dioxide and hydrogen as the main constituents is obtained by gasification of MSW. This gas could be stored and used when required for power generation [41]. The waste-feeding rate varies as employed but is usually about 50–150 kg/h and its efficiency is about 70–80%. Gasification utilizing plasma with steam results in a hydrogen and carbon dioxide rich "synthetic" gas commonly known as syngas. Gasification in the presence of small amounts of air produces syngas which is rich in nitrogen but of low quality in terms of calorific value, and with oxygen produces a high quality fuel mixture of carbon monoxide and hydrogen.

This process utilizing high temperature plasma requires electrical power to produce the hot chamber facilitating gasification. As discussed the main product obtained is syngas, along with slag as a by-product. The process breaks down nearly all the feedstock components into their elemental forms excluding the radioactive materials which facilitates into decomposition of toxic compounds into harmless chemical elements, which is also one of the main advantages offering in comparison with the conventional methods of waste disposal [41].

There are usually five processes for elimination of tars: thermal cracking, catalytic cracking, mechanical methods (scrubber, filter, cyclone, and electrostatic precipitator), self-modifications (operating parameters) and plasma methods [42, 43]. Fabry et al. [44] has proposed an overview of gasification by thermal plasma process in converting waste to energy. All aspects regarding gasification, like chemical reactions, configuration of main reactor, tar content and other operation conditions have been discussed in details. Comparison with other waste gasification processes (using DC or AC plasma torches) has been done successfully. It also gives a review of various torches used in these processes. Shiota et al. [45] investigated the emission of particulate matter (PM) from gasification in Japan. The size distribution and removal efficiency of PM was investigated from 3 different gasification furnaces. A nine-stage cascade was used for this purpose. It was showed that most PM was fine particles and the coarse particles were removed efficiently. The removal efficiency of PM and PM2.5 were 99.96 and 99.99% in plants BF-SCR and BF-wet scrubber-SCR-BF system. Cao et al. [46] through thermodynamic analysis assessed the hydrogen production from superficial water gasification of diosgenin solid waste. Aspen plus software was used for the analysis and is based on the principle of Gibbs free energy. K_2CO_3 and black liquor was used to catalyse the gasification and the addition of black liquor significantly improved the hydrogen yield.

Couto et al. [47] evaluated the MSW gasification in Portugal thermodynamically. Energy and exergy values were strictly investigated in order to find the result. It was observed that tar content energy value decreased considerably (almost 80%) with the increase in temperature (900 °C). Pandey et al. [48] predicted the lower heating value of gas, other gasification products such as tar, entrained char and syngas with the help of multi-layer feed forward neural networks. Rigorous study was carried out in choosing the no. of hidden layers, no. of neurons in the hidden layers and activation function in a network. 9 input and 3 output parameters were taken into consideration for the same. This showed that the performance of fluidized bed gasifier can be predicted by artificial neural network. Nielsen et al. [49] invented a plasma-assisted waste gasification system and the reactor has 3 zones namely a bottom zone, a middle zone and a top zone. The bottom zone is used for melting the waste reaction residue and forming a slag pool. The middle layer is used to make syngas from the waste and the top zone has a plasma torch thus controlling the temperature of syngas. The syngas produced by gasification of MSW in gas turbine or reciprocating engine has a higher efficiency than direct combustion producing thermal energy in steam engine [50].

3 Potential of MSW in Energy Production for the Society

Citing the tremendous increase in the MSW generation potential per capita, is a major concern for our mother Earth and our sustainability. As per [51] almost one billion tons rise in MSW generation is expected by the year 2025. By widening our research focus towards enhancement in waste disposal technology in order to extract energy from waste in some form which could benefit the society, could mitigate several waste management issues. An estimated potential of energy generation from MSW in Malaysia crosses over 600 kW per day with waste generation of 1500 tons having 2200 kcal/kg heating rate [52]. Analysing the market for thermal treatment of wastes, incineration could generate electricity from waste treatment products at an efficiency of merely 20% whereas gasification, with WTE conversion efficiency of 34% leads the role in producing electricity from MSW [53]. This could pave the way for sustainable adaptation of WTE technologies for alleviation of fossil fuel issues.

4 Conclusion

Tremendous amount of waste has already been disposed of in unscientific manner causing perilous concerns to the society and to the environment raising global warming issues. The landfills piled up are now burden not only to the society, but also to the governments dealing to dispose it off. Concerns over waste management problems should now be reformed into technical solutions for generating useful energy, from the gigantic landfills and the impending waste to be collected. Involvement of chemical, mechanical and environmental engineering is need of the hour. WTE products as well as by-products are capable of providing energy back to the society. Technologies like composting, bio-methanation and pyrolysis not only dispose of the wastes but also produce fruits for the society. Analysing the quantity of MSW accumulated as well as the anticipated wastes, it can promise to contribute to a greater extent in energy generation scenario at national as well as international status.

References

1. Nishikawa, H., Ibe, M., Tanaka, M., Takemoto, T., & Ushio, M. (2006). Effect of DC steam plasma on gasifying carbonized waste. *Vacuum, 80,* 1311–1315.
2. Morcos, V. (1989). Energy recovery from municipal solid waste incineration. *Heat Recover System CHP, 9,* 115–1126.
3. Ansbakov, A. S., Faleev, V. A., & Kezevich, D. D. (2002). Domestic waste plasma gasification technology and its comparison with ordinary one burning on the final products. In *Ecology, electrotechnology and waste processing*; KORUS (pp. 211–213).
4. Lyubina, Y. L., & Suris, A. L. (1999). Thermodynamic model of the plasma gasification of organic solid waste. *Chemical and Petroleum Engineering, 35,* 38–40.

5. Central Pollution Control Board, India. Waste management programmes annual report 2015–16. http://www.cpcb.nic.in/Municipal_Solid_Waste.php. Accessed November 16, 2017.

6. Asansol Municipal Corporation, India. (2017). Sanitation Department. http://www.cpcb.nic.in/Municipal_Solid_Waste.php. Accessed November 16, 2017.

7. Nema, S. K., & Ganeshprasad, K. S. (2002). Plasma pyrolysis of medical waste. *Current Science, 83,* 271–278.

8. Dave, P.N., & Joshi, A. K. (2010). Plasma pyrolysis and gasification of plastic waste - a review. *Journal of Scientific & Industrial Research, 69,* 177-179.

9. Punčochář, M., Ruj, B., & Chatterjee, P. K. (2012). Development of process for disposal of plastic waste using plasma pyrolysis technology and option for energy recovery. *Procedia Engineering, 42,* 420–430.

10. Ojha, A., Reuben, A. C., & Sharma, D. (2012). Solid waste management in developing countries through plasma arc gasification-An alternative approach. *APCBEE Procedia, 1,* 193–198.

11. Detailed project report (DPR) for municipal solid waste disposal through incineration process by Pondicherry Urban Development Agency—Local Administration Department on November 2014.

12. Detailed project report (DPR) on municipal solid waste management for Vishakhapatnam, prepared and submitted by Feedback Infra Private Limited in JV with Eco save Systems Pvt Ltd on September 2015.

13. California's Department of Resources Recycling and Recovery, US. (2007). Statewise waste generated, diverted and disposed. http://www.ciwmb.ca.gov/LGCentral/Rates/Graphs/RateTable.htm. Accessed November 16, 2017.

14. Williams, R., Jenkins, B., & Nguyen, D. (2003). *Solid waste conversion: A review and database of current and emerging technologies* (pp. 1–129). Final Report: Department of Biology.

15. Williams, R. B., Jenkins, B. M., & Kaffka, S. (2013). California BC. An Assessment of Biomass Resources in California, Data. Contract Report to CEC PIER Contract 500-11-020 2015:1–155.

16. Leary, M. (2004). Contractor's Report to the Board Landfill Facility Compliance Study Task 6 Report—Review of MSW Landfill Regulations from Selected States and Countries.

17. Commission of the European Communities. (2006). Green Paper on a future Maritime Policy for the EU presented by the commission; SEC, p. 689.

18. Municipal Solid Waste in the United States. (2005). Facts and figures. US Environmental Protection Agency, municipal and industrial solid waste division; 2006: PA530-R-06-011.

19. UN-HABITAT. (2009). Solid waste management in the world's cities. United Nations Human Settlements Programme.

20. Sharholy, M., Ahmad, K., Mahmood, G., & Trivedi, R. C. (2008). Municipal solid waste management in Indian cities—A review. *Waste Management, 28,* 459–467.

21. Cheng, H., & Hu, Y. (2010). Municipal solid waste (MSW) as a renewable source of energy: Current and future practices in China. *Bioresource Technology, 101,* 3816–3824.

22. Hamer, G. (2003). Solid waste treatment and disposal: Effects on public health and environmental safety. *Biotechnology Advancement, 22,* 71–79.

23. Freudenrich, C. Ph.D. How Landfills Work. 16 October 2000. How stuff works. http://science.howstuffworks.com/environmental/green-science/landfill.html; 2017. Accessed June 9, 2017.

24. Szabó, S., Bódis, K., Kougias, I., Moner-Girona, M., Jäger-Waldau, A., Barton, G., et al. (2017). A methodology for maximizing the benefits of solar landfills on closed sites. *Renewable and Sustainable Energy Reviews, 76,* 1291–1300.

25. Hartmann, B., Török, S., Börcsök, E., & Oláhné, Groma V. (2014). Multi-objective method for energy purpose redevelopment of brownfield sites. *Journal Clean Production, 82,* 202–212.

26. Mønster, J., Samuelsson, J., Kjeldsen, P., & Scheutz, C. (2015). Quantification of methane emissions from 15 Danish landfills using the mobile tracer dispersion method. *Waste Management, 35,* 177–186.

27. Vázquez, M. A., & Soto, M. (2017). The efficiency of home composting programmes and compost quality. *Waste Management, 64,* 39–50.

28. Yang, D., Zheng, L., Song, W., Chen, S., & Zhang, Y. (2012). Evaluation indexes and methods for water quality in ocean dumping areas. *Procedia Environment Science, 16,* 112–117.

29. Simonini, R., Ansaloni, I., Cavallini, F., Graziosi, F., Iotti, M., Massamba N'Siala, G., et al. (2005). Effects of long-term dumping of harbor-dredged material on macrozoobenthos at four disposal sites along the Emilia-Romagna coast (Northern Adriatic Sea, Italy). *Marine Pollution Bulletin, 50,* 1595–1605.
30. Zheng, L., Cui, W., Song, W., Qu, L., Yuan, Y., & Yang, D. (2012). The biological effects of the marine dumping on mollusks. *Procedia Environment Science, 16,* 118–124.
31. Wang, Y., Cheng, K., Wu, W., Tian, H., Yi, P., Zhi, G., et al. (2017). Atmospheric emissions of typical toxic heavy metals from open burning of municipal solid waste in China. *Atmosphere Environment, 152,* 6–15.
32. Szijjarto, A., & Hungerb, K. (2017). Optimized energy use through systematic short—Term management of industrial waste incineration. *Computers and Chemical Engineering, 1354*(17), 30151-301515.
33. Nabavi-pelesaraei, A., Bayat, R., Hosseinzadeh, H., Afrasyabi, H., & Chau, K. (2017). Modelling of energy consumption and environmental life cycle assessment for incineration and landfill systems of municipal solid waste management—A case study in Tehran Metropolis of Iran. *Journal of Clean Production, 6526*(17), 30194-4.
34. Zhou, X., Zhou, M., Wu, X., Han, Y., Geng, J., Wang, T., et al. (2017). Reductive solidification/stabilization of chromate in municipal solid waste incineration fly ash by ascorbic acid and blast furnace slag. *Chemosphere, 182,* 76–84.
35. Hwang, K. L., Choi, S. M., Kim, M. K., Heo, J. B., & Zoh, K. D. (2017). Emission of greenhouse gases from waste incineration in Korea. *Journal of Environment Management, 196,* 710–718.
36. Gradus, R. H. J. M., Nillesen, P. H. L., Dijkgraaf, E., & van Koppen, R. J. (2017). A cost-effectiveness analysis for incineration or recycling of Dutch household plastic waste. *Ecological Economics, 135,* 22–28.
37. Song, J., Sun, Y., & Jin, L. (2017). PESTEL analysis of the development of the waste-to-energy incineration industry in China. *Renewable and Sustainable Energy Reviews, 80,* 276–289.
38. Ansah, E., Wang, L., & Shahbazi, A. (2016). Thermogravimetric and calorimetric characteristics during co-pyrolysis of municipal solid waste components. *Waste Management* (article in press).
39. Gunasee, S. D., Carrier, M., Gorgens, J. F., & Mohee, R. (2016). synergistic effects using TGA-MS. *Journal of Analytical and Applied Pyrolysis,* 1–12.
40. Uro, Vlv., Ri, K., Lq, D., Qtldqj, D. X., Hl, K., Lq, L. D. R. X., et al. (2017). JHQ IUHH & LUFXPVWDQFH *105,* 1255–1262. https://doi.org/10.1016/j.egypro.2017.03.442.
41. Gray, L. (2014). Plasma gasification as a viable waste-to- energy treatment of municipal solid waste. MANE-6960; Solid and Hazardous Waste Prevention and Control Engineering.
42. He, J., & Zhang, W. (2011). Review of syngas production via biomass. *Renewable and Sustainable Energy Reviews, 15,* 482–492.
43. Han, J., & Kim, H. (2008). The reduction and control technology of tar during biomass gasification/pyrolysis: An overview *Renewable and Sustainable Energy Reviews, 12,* 397–416.
44. Fabry, F., Rehmet, C., Rohani, V., & Fulcheri, L. (2013). Waste gasification by thermal plasma: A review. *Waste Biomass Valorisation, 3,* 421–439 (Springer).
45. Shiota, K., Tsujimoto, Y., Takaoka, M., Oshita, K., & Fujimori, T. (2017). Emission of particulate matter from gasification and melting furnace for municipal solid waste in Japan. *Journal of Environment Chemical Engineering, 17,* 30102–1.
46. Cao, W., Cao, C., Guo, L., & Jin, H. (2017). Gasification of diosgenin solid waste for hydrogen production in supercritical water. *International Journal of Hydrogen Energy,* 1–10.
47. Dinis, N. C., Bruno, V. S., & Rouboa, A. (2016). Thermodynamic evaluation of Portuguese municipal solid waste gasification. *Journal of Cleaner Production, 139,* 622–635.
48. Shankar, D. P., Das, S., Pan, I., Leahy, J. J., & Kwapinski, W. (2016). Artificial neural network based modelling approach for municipal solid waste gasification in a fluidized bed reactor. *Waste Management* (article in press).
49. Nielsen, M. C. (19). United States (12) 2009;1. Patent application US20090064581A1.
50. Zhao, P., Ni, G., Jiang, Y., Chen, L., Chen, M., & Meng, Y. (2010). Destruction of inorganic municipal solid waste incinerator fly ash in a DC arc plasma furnace. *Journal of Hazardous Materials, 181,* 580–585.

51. Kosuke, K., & Tomohiro, T. (2014). Revisiting estimates of municipal solid waste generation per capita and their reliability. *Journal of Material Cycles and Waste Management, 2014–2015.*
52. Ivapalan, K., Muhd, Y., Kamaruzzaman, S., & Abdul, S. (2003). Energy potential from municipal solid waste in Malaysia. *Renewable Energy, 29,* 559–567.
53. Murphya, D., & McKeogh, E. (2004). Technical, economic and environmental analysis of energy production from municipal solid waste. *Renewable Energy, 29,* 1043–1057.

Design of a Decentralised Biogas Reactor

R. Hoque, R. L. Kshetry, S. Maity, R. Dalui and A. Khamaru

Abstract With a population of 4,496,694, Kolkata is the 3rd largest megacity in India and the 20th largest in the world. But this growth has its consequences. Kolkata metropolitan area produces approximately 4000 megatons of waste every day, of which solid wastes comprise approximately 3932 megatons. These wastes are handled by the Kolkata Municipal Corporation (KMC). In the wake of the energy crisis and the depletion of fossil fuel reserves, the need to switch to non-conventional sources of energy is looming. The most notable among these is biogas, which can theoretically generate 1.4 times the annual global power requirement. Despite several works having been done in this field, biogas is still quite an untapped energy source. This paper explores the possibilities of using biogas for cooking purposes as an alternative fuel of liquefied petroleum gas (LPG), and proposes the design and construction of a decentralized biogas reactor for a small apartment building of 10–15 families with an assumed four members per family. Calculations and estimations showed that the biogas thus generated by the suggested two-stage process from solid wastes produced by the families would be sufficient to serve the purpose of substituting the cooking gas of all families residing in that building, thereby solving the dual problem of solid waste management and providing a renewable source of energy as a cheap cooking fuel.

R. Hoque (✉) · R. L. Kshetry · S. Maity · R. Dalui
Department of Mechanical Engineering, Institute of Engineering & Management,
Kolkata 700091, West Bengal, India
e-mail: rh2974@columbia.edu

R. L. Kshetry
e-mail: rlkshetry95@gmail.com

S. Maity
e-mail: sagnik.m21@gmail.com

R. Dalui
e-mail: ritabratadalui@gmail.com

A. Khamaru
Centre for Material Science and Engineering, National Institute of Technology,
Hamirpur 177005, Himachal Pradesh, India
e-mail: anindya55555@gmail.com

© Springer Nature Singapore Pte Ltd. 2020
S. K. Ghosh (ed.), *Energy Recovery Processes from Wastes*,
https://doi.org/10.1007/978-981-32-9228-4_14

Keywords Municipal solid wastes · Biomass energy · Biogas generation ·
Decentralised biogas reactor · Anaerobic fermentation

1 Introduction

Joshi et al. [1] found that Indian urban areas produce municipal solid wastes (MSW)
at a rate of 0.5–0.7 kg/capita/day and 75% of it is dumped in unsanitary landfills
or open dumpsites. They claimed that rapid population growth in India, especially
in urban areas, is the main cause of hefty increase of MSW in India, at a rate of
1.33% per capita per year. They proposed a "Decentralized MSW Treatment" plan
which would see the setting up of compact, decentralized composting units. Kumar
and Kaushal [2] compared different methods of municipal solid waste management
(MSWM) practiced in India.

They found that open dumping or unsanitary landfill is still the most common
method of waste management in the country. After a detailed case study, they concluded that material recovery (recycling) and composting are the best methods of
MSWM, in terms of least damage caused to the environment.

Das et al. [3] did a case study on municipal solid waste management (MSWM) in
Kolkata. The survey was carried out on 41 municipal towns in the Kolkata metropolitan area, results of which echoed Joshi's [1] findings. Das and Bhattacharyya [4]
gave an overview of solid waste management in Kolkata. They studied the practiced
methods of waste collection and recommended improvements in the transportation
of wastes. De and Debnath [5] selected a garbage disposable area of Kolkata in Garia.
Nearby households (within 500 m from the waste disposable land) were randomly
selected and a survey was conducted on the consequence of garbage disposal on
the health of the residents, taking note of perception and awareness about garbage
disposal practices. The results of the study clearly indicate that in the absence of a
basic facility of collection of waste from source, citizens were forced to dump waste
on the open spaces, streets, drains and water bodies in the surrounding area, creating
unsanitary conditions.

Municipal solid waste (MSW) is defined as the waste comprising of discarded
items of everyday use, such as food wastes, paper, cardboard, textiles, plastics, glass,
metal, etc. Solid wastes basically comprise of organic wastes (food scraps, yard
wastes, leaves, grass, wood, etc.), paper (cardboard, newspapers, books, magazines,
paper bags, paper cups, etc.), glass (bottles, broken light bulbs, broken window
fragments, etc.), plastics (bottles, packaging material, containers, cups, etc.), metal
(cans, foils, parts of appliances, etc.) and other wastes like textiles, leather, e-wastes,
rubber, ash, etc.

In India solid wastes are treated in eight different ways namely:

1. Composting
2. RDF (Refuse derived fuel) production
3. Pyrolysis
4. LFG (Landfill gas) recovery

5. Sanitary landfill
6. Uncontrolled dumping
7. Biomethanation
8. Incineration

Of these methods only composting and uncontrolled dumping are prevalent in Kolkata.

Rapid industrial, economic and infrastructural growth in India over the past few decades have led to skyrocketing urban population. Presently, urban population accounts for as much one-third of the entire population of the country. Currently, Indian urban areas produce municipal solid wastes (MSW) at about 0.5–0.7 kg per capita per day [2], which is quite high. Data on the situation of MSWM in India is tabulated below (Fig. 1; Tables 1 and 2).

It is estimated that MSW generated from Indian urban areas is increasing at an alarming rate of 1.33% per capita per year, and Kolkata is no exception. As much as 75% of the solid waste generated in India is dumped in unscientifically managed landfills or open dumpsites [1]. The fields of Dhapa in Kolkata are an example of such open dumpsite. The waste such treated ultimately undergoes natural redox reaction by virtue of aerobic bacteria. However, such dumpsites serve as spawning ground for disease-causing germs, viruses and bacteria. Moreover, poisonous leachates from decomposed waste also pollute the ground.

In some places, if the water table is high, the leachates seep into the water and pollute it. Drinking this polluted water can lead to serious health hazards. Besides, unsanitary landfills emit foul odour, and also affect the aesthetic beauty of the locality.

Industrial development and population growth have also led to a surge in the global demand for energy in recent years. Consequently, energy reserves are depleting

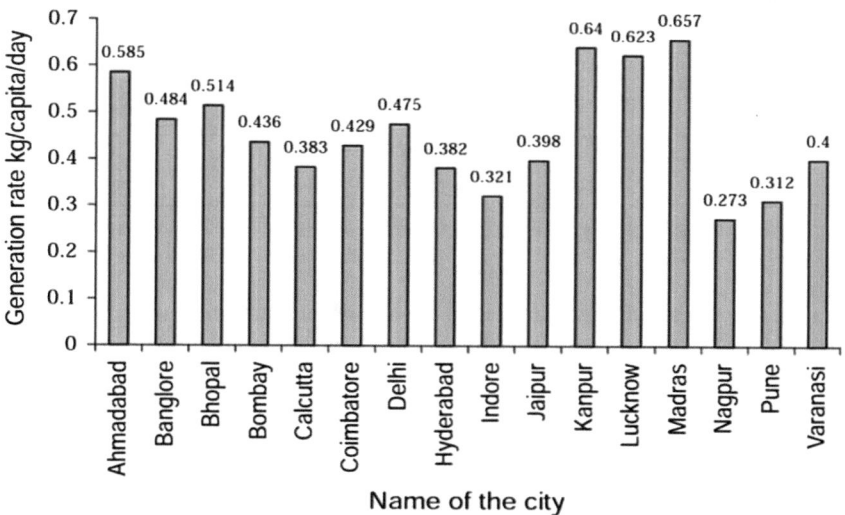

Fig. 1 Per capita generation rate of MSW for Indian cities (CPCB 2004)

Table 1 Municipal solid waste generation rates in different states in India

S. No.	Name of the state	No. of cities	Municipal population	Municipal solid waste (t/day)	Per capita generated (kg/day)
1	Andhra Pradesh	32	10,845,907	3943	0.364
2	Assam	4	878,310	196	0.223
3	Bihar	17	5,278,361	1479	0.280
4	Gujarat	21	8,443,962	3805	0.451
5	Haryana	12	2,254,353	623	0.276
6	Himachal Pradesh	1	82,054	35	0.427
7	Karnataka	21	8,283,498	3118	0.376
8	Kerala	146	3,107,358	1220	0.393
9	Madhya Pradesh	23	7,225,833	2286	0.316
10	Maharashtra	27	22,727,186	8589	0.378
11	Manipur	1	198,535	40	0.201
12	Meghalaya	1	223,366	35	0.157
13	Mizoram	1	155,240	46	0.296
14	Orissa	7	1,766,021	646	0.366
15	Punjab	10	3,209,903	1001	0.312
16	Rajasthan	14	4,979,301	1768	0.355
17	Tamil Nadu	25	10,745,773	5021	0.467
18	Tripura	1	157,358	33	0.210
19	Uttar Pradesh	41	14,480,479	5515	0.381
20	West Bengal	23	13,943,445	4475	0.321
21	Chandigarh	1	504,094	200	0.397
22	Delhi	1	8,419,084	4000	0.475
23	Pondicherry	1	203,065	60	0.295
		299	128,113,865	48,134	0.376

Source: Status of MSW generation, collection, treatment and disposal in class-I cities (CPCB 2000)

all over the earth. This problem is aggravated by the lack of long-term planning by the world's governments. Despite various efforts by the United Nations, non-conventional sources of energy are far from being used regularly for a source of stable power. The major reason for this is the dearth of infrastructure to harness the energy in a sustainable way.

One of the most notable sources of energy among the non-conventional ones is biomass energy. Biomass is a general term for any living material. Taken together, the earth's biomass represents an enormous source of energy. The energy reserve

Table 2 Physical characteristics of MSW in Indian metrocities

Characteristics (% by weight)

Name of metrocity	Paper	Textile	Leather	Plastic	Metals	Glass	Ash, fine earth and others	Compostable matter
Ahmedabad	6.0	1.0	–	3.0	–	–	50.0	40.00
Bangalore	8.0	5.0	–	6.0	3.0	6.0	27.0	45.00
Bhopal	10.0	5.0	2.0	2.0	–	1.0	35.0	45.00
Mumbai	10.0	3.6	0.2	2.0	–	0.2	44.0	40.00
Calcutta	10.0	3.0	1.0	8.0	–	3.0	35.0	40.00
Coimbatore	5.0	9.0	–	1.0	–	–	50.0	35.00
Delhi	6.6	4.0	0.6	1.5	2.5	1.2	51.5	31.78
Hyderabad	7.0	1.7	–	1.3	–	–	50.0	40.00
Indore	5.0	2.0	–	1.0	–	–	49.0	43.00
Jaipur	6.0	2.0	–	1.0	–	2.0	47.0	42.00
Kanpur	5.0	1.0	5.0	1.5	–	–	52.5	40.00
Kochi	4.9	–	–	I.I	–	–	36.0	58.00
Lucknow	4.0	2.0	–	4.0	1.0	–	49.0	40.00
Ludhiana	3.0	5.0	–	3.0	–	–	30.0	40.00
Madras	10.0	5.0	5.0	3.0	–	–	33.0	44.00
Madurai	5.0	1.0	–	3.0	–	–	46.0	45.00
Nagpur	4.5	7.0	1.9	1.25	0.35	1.2	53.4	30.40
Patna	4.0	5.0	2.0	6.0	1.0	2.0	35.0	45.00
Pune	5.0	–	–	5.0	–	10.0	15.0	55.00
Surat	4.0	5.0	–	3.0	–	3.0	45.0	40.00
Vadodara	4.0	–	–	7.0	–	–	49.0	40.00
Varanasi	3.0	4.0	–	10.0	–	–	35.0	48.00
Visakhapatnam	0	2.0	–	5.0	–	5.0	50	35.00
Average	5.7	3.5	0.8	3.9	1.9	2.1	40.3	41.80

Source: Status of solid waste generation, collection, treatment and disposal in metrocities (CPCB 2000)

per ton of biomass is approximately 1500–3000 kWh. Theoretically, biomass can generate almost 1.4 times the earth's annual energy consumption of 150×10^3 TW-h [6]. Compared to solar energy, biomass is more efficient and can generate more energy with better economy. The annual world energy requirement is 17.7 TW-h, while the energy that can be generated from biogas can range from 16 to as much 33 TW-h, depending on the method used, substrate used, process parameters, and process efficiency. So at this junction, biogas generation is considered to be a suitable solution to the dual problem of solid waste management as also depletion of fossil fuel reserves.

In this light, Bhol et al. [7] reviewed the current state of biogas digesters in India and claimed that biogas generation is an extremely ignored way when it comes to reduction of greenhouse gas emissions and providing a better waste disposal method for organic waste. If applied properly and under suitable circumstances, biogas energy can be harnessed to completely replace wood, coal or LPG as the prominent cooking fuel in at least the rural and semi-urban sectors. They suggested that biogas, being cheap, may also be used for electricity generation in the domestic as also community level in villages. They discussed the construction, design and operating principle of various biogas digesters currently used in India—namely, the KVIC (Khadi & Village Industries Commission of India) model, the Janata model and the Deenabandhu model. Karve [8] also surveyed the general status of biogas in India. He claimed that despite there are 2–3 million biogas plants in working condition in India, but hardly 2% rural families use biogas, citing that the main reason for this was inefficient plants, which, due to their improper design, end up making biogas costlier than its alternatives—almost 1.3 times as expensive as LPG (based on an estimated daily use of 500 g LPG by each family). Pohekar et al. [9] also researched about the plausibility of using alternative cooking fuel in India. They claimed that, out of the total energy consumption in India, cooking energy requirements for industrial and domestic purposes both, in rural and urban areas combined, account for a significant 36%. They hypothesized that almost 90% of the household energy consumption in a middle-class Indian urban family is for cooking only. In rural areas, 75% of energy requirements of a household are owed to wood (timber) and agriculture waste; the rest is met by kerosene and liquefied petroleum gas (LPG), as stated by Reddy [10]. According to a series of studies by the National Sample Survey Organization (NSSO) [11], the type of fuel used by a particular household or family would depend upon its net income. A report by MNES (Ministry of Non-Conventional Energy Sources) published in 2003 [12] claims that India has a latent potential for construction and successful functioning of 12 million biogas plants, but several constraints like disinterest from governments, dearth of technology, lack of funding, etc. have proven deterrent to this development. However, the alarming depletion of fossil fuel reserves throughout the world has helped indirectly for biogas and biofuel to be thrown under the spotlight. The National Programme on Biogas Development (NPBD) was started by the MNES in 1981, and consequently the number of installed biogas plants in India by the end of March 2003 had risen to 3.523 million [13] (according to Winrock International).

Satpathy et al. [14] studied the influence of various substrates on the microbial communities in batch type biogas reactors working with maize and maize silage as substrates. They underlined the importance of lactate in biogas reactors, and further modified their concept with the help of the experimental findings in another paper [15].

Biomass still remains an untapped source of energy, despite having several advantages, such as renewability, lower pollution than fossil fuels, viable solution to waste management problems, indigenousness of source, non-dependence on weather or atmospheric conditions, and providing opportunities for economic development.

Coming back to the problem of MSWM in Kolkata, one of the gravest issues that can be addressed in this regard is source segregation. Few households do promote separate collection of organic and inorganic wastes but in most cases, the Kolkata Municipal Corporation does not provide separate collection and disposal avenues. So the effort given in source segregation becomes worthless, as the segregated waste is mixed again in collection vehicles or disposal grounds. Another major issue is the problem of actual collection of wastes. In Kolkata, the waste is collected door-to-door by workers of the KMC. However, this covers only 65% of the household in the urban area of Kolkata [16]. A large fraction of the household wastes is thrown out haphazardly. Also, due to narrow lanes and streets, collection vehicles cannot reach everywhere. Those places have to depend on sweepers for waste collection. Due to large scale deployment of manpower, the labour cost becomes high. Another problem related to this is that all the wastes are openly dumped at the dumping grounds of Dhapa without any form of treatment. Even after the use of compactors, if the wastes are openly dumped, they still cause pollution, and the use of compactors is totally negated.

This paper explores the possibilities of using biogas as an alternative fuel for cooking purposes instead of liquefied petroleum gas (LPG) thereby simultaneously solving the problem of waste management as the organic solid waste is being used up by the family itself—while at the same time reducing the problem of energy generation and pollution. The use of this process also reduces the volume and thereby the cost of handling waste.

2 Methodology

In the proposed model of the biogas reactor we aimed to utilize anaerobic fermentation process to design a small-sized, decentralized, portable, fixed dome biogas reactor for application in the urban environment. Indian urban areas produce municipal solid waste at a rate of 0.5–0.7 kg/capita/day and major portion of it organic solid waste which can be used to generate biogas in a biogas reactor. In anaerobic fermentation method biogas is produced from wet biomass with about 90–95% water content by the action of anaerobic bacteria. These bacteria live and grow in the absence of oxygen and they derive the needed oxygen by decomposing the biomass. The process is favoured by wet, warm and dark conditions. The biochemical process proceeds in three stages:

Stage 1: The original organic matter containing complex compounds like carbohydrate, protein, fats etc. are broken through the influence of water to simple water-soluble compounds. The polymers are reduced to monomers. The process takes about a day at 25 °C in an active digester.

Stage 2: The micro-organisms of anaerobic and facultative groups, together known as acid formers, produce mainly acetic and propionic acids. This stage also takes about one day at 25 °C. Much of CO_2 is released in this stage.

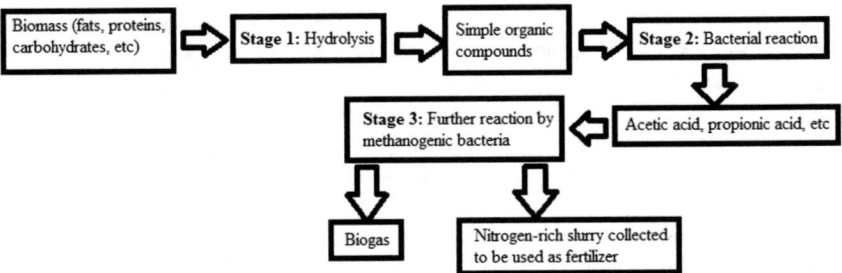

Fig. 2 Flowchart to describe anaerobic fermentation process

Stage 3: Anaerobic bacteria, also known as methane formers slowly digest the products available from the second stage to produce methane, CO_2, small amount of hydrogen and trace amount of other gases. The process takes about two weeks of time to complete at 25 °C. This stage is carried out strictly by the action of anaerobic bacteria.

Our design is suited for midrise apartments housing ten to fifteen families with four members per family generating about 12–20 kg of solid waste. The organic solid waste which has to be segregated at the source (in each respective house) will be collected in a container on the roof of the apartment. Water will be sprayed on top of the waste continuously in order to maintain the water content required to carry out the biogas generation process. The water and waste mixture will then be fed through a PVC pipe into a dome, which will be placed about 6–7 feet above the ground on an elevated platform. This dome would contain methanogenic bacteria. Stage 1 and stage 2, as described previously, will be carried out within this dome. A PRV (pressure regulating valve) will release the CO_2 produced during stage 2 into the atmosphere. The substrate generated at the end of stage 2 will be directed to another dome placed at a height of 1 foot from the ground where stage 3 will transpire. The methane generated in this stage will be directed through a PVC pipe to a sealed container through a PRV from where it will be distributed to the individual families.

A flashback arrestor will have to be provided to prevent accidental combustion and reverse propagation of flame front. Nutrients such as soluble nitrogen compounds will remain available in the solution or slurry left in the tank, and can provide excellent fertilizer and humus. The solution will be drained from the bottom of the second dome and can be used as a highly effective fertilizer.

The processes can be described in the form of a simple flowchart (Fig. 2).

The schematic diagram of the proposed full-scale reactor system is shown in Fig. 3.

3 Experimental Setup and Procedure

To practically implement our idea, we constructed a small prototype.

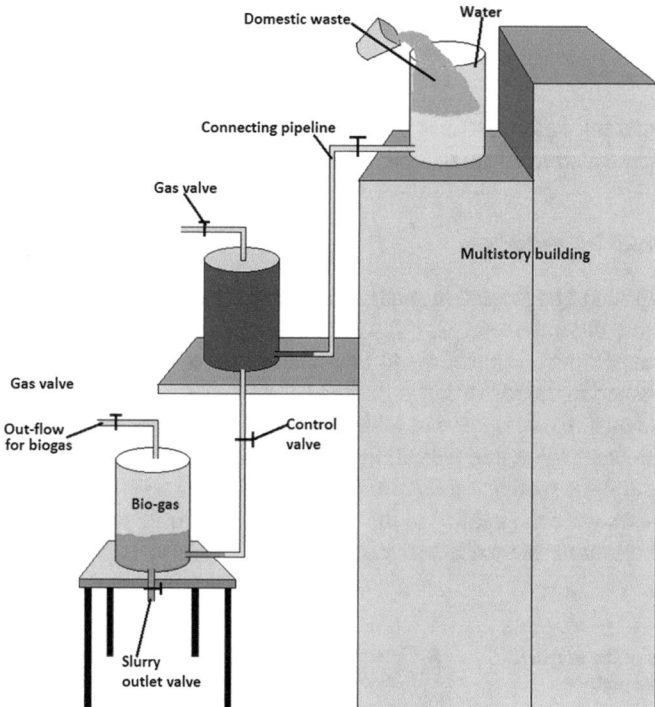

Fig. 3 Schematic diagram of full-scale proposed model

Raw materials required:

- Hard polyethylene water drums (to serve as the reactor dome as well as storage tank)
- Mild steel L-angles, nuts & bolts (to construct the frame)
- Hard PVC pipes, butterfly valve (to allow flow of slurry)
- Gas pipes, pipe bends & fittings, nozzles, flashback arrestor (for supply of generated gas)
- GI sheet, resistive foam (polyurethane), adhesive tape, adhesive resin (Araldite), sealant (M-seal).

A brief idea about the manufacturing procedure:

- The main framework was constructed using L-angles attached together using nuts &bolts.
- The ends of each drum were cut off as per required dimension to accommodate the pipe fittings.
- The drums were fitted at their respective positions and the PVC pipes were attached.
- The joints were sealed using M-seal, adhesive tape and Araldite.
- The gas pipe was attached to the bottom-most tank and sealing was properly done.

- Flashback arrestor and nozzle were attached.

Cost Estimation:

The proposed full-scale reactor model would be a low cost one. In correspondence with that, the constructed small-scale reactor was also quite cheap (Fig. 4; Tables 3 and 4).

Experimental Procedure:

- The setup had to be located in shade away from direct sunlight in such a place that temperature did not exceed 25 °C.
- For the experimental procedure to be undertaken, fresh cow dung was collected and deposited in the top drum.
- Requisite quantity of water was added and stirred properly.
- The drum was sealed and was allowed to rest for the retention period of 3 days. This allowed the methanogenic bacteria in the dung to be activated.
- Food wastes were then added to the cow dung and sealing was properly done.
- The entire setup was then left to rest.

Fig. 4 View of the prepared small-scale laboratory prototype

Table 3 Dimensions of the constructed laboratory model reactor

Part	Parameter	Dimension (cm)
Drum (Roughly assumed as cylindrical)	Mean outer circumference	83
	Mean inner circumference	82
	Mean thickness	0.5
	Mean inner diameter	13
	Height (Base to neck)	44
	Actual available height inside	43
PVC pipe (For slurry transport)	Outer circumference	37
	Inner circumference	36
	Thickness	0.5
	Inner diameter	6
	Length between the tanks	67
Gas pipe	Inner circumference	7
	Inner diameter	1.2

Table 4 Cost estimation of the prototype

Products/components	Quantity	Price (INR)
PVC containers	3	150/-
Gas pipe (0.5 inch diameter)	Length as required	300/-
Stop valve	2	150/-
Gas regulator	2	120/-
PVC pipe (6 inch diameter)	Length as required	500/-
Flashback arrestor	1	150/-
L-angles	48 feet	1500/-
Methanogenic bacteria		0
Other miscellaneous costs		200/-
Total		3070/-

4 Results and Discussions

The experiment proved that biogas was successfully generated, albeit not in a large amount. The quantifiable results are listed in Table 5.

The experimental data correlates with the estimated amount of biogas that could be generated from the amount of waste input. The deviation of the actual amount from the theoretical value could be attributed to the following factors:

- Insufficient amount of carbohydrates in the food waste sample
- Inadequate quantity of bacteria generated from the cow dung

Table 5 Results of the experiment

Amount of solid waste used as substance	2 kg
Amount of cow dung used	350 g
Amount of water added	3.5 l
Gas generated	150 ml
Available energy	4950 J

- Unfavourable reaction conditions
- Varying ambient temperature conditions
- Improper amount of water supplied
- Improper mixing and stirring of the waste with water
- Insufficient retention period
- Short reaction time
- Inability to attain perfectly anaerobic condition

Despite these factors, some quantity of biogas was definitely generated, as was proven by testing it for combustion. Comparatively, it can be noted that Farouk et al. [17] in their research project, utilized a total volume of 16.5 l of 10% DM fresh cow dung manure mixture in a laboratory-scale digester (which was tightly closed to allow anaerobic digestion), and thereby obtained 17.5 l of biogas over a span of 38 days, i.e. approximately 460 ml/day. The starting temperature of the experiment was low at 21 °C but increased up to 30 °C at the end of the experiment. Performing the same experiment using dry cow dung they obtained 15 l of biogas in 15 days, i.e. 1 l/day. The operating temperature during the experiment was constant within the range of 37–38 °C. In comparison, Samuchit Enviro Tech, a social enterprise which produces and supplies biogas commercially in Asia & Africa, claim to generate approx. 430 ml of biogas daily in their domestic balcony model biogas reactors from 1 kg of kitchen wastes.

Biogas itself burns with a slightly bluish inner core with a reddish-yellow flame. The extent of blueness in the flame is governed by the amount of methane in the gas. Greater is the amount of methane, more prominent blue coloured core is observed during combustion. In general, biogas is comprised of 55–65% methane and 30–35% carbon dioxide, the remaining being trace amounts of ethane, hydrogen sulphide, ammonia, carbon monoxide, nitrogen, oxygen, and other hydrocarbons. This mixture depends on the composition of the substrate which is being decomposed for this purpose, chemical analysis of the food waste which is being decomposed, and also of the slurry which is left after decomposition, should be done in laboratory. The temperature, pH value, BOD (biological oxygen demand), COD (chemical oxygen demand), etc. should be measured, which we unfortunately could not do due to various constraints.

5 Conclusion

An analysis of the current situation of MSWM in Kolkata shows some positive accolades under the belt of the Kolkata Municipal Corporation. KMC now handles approximately 4000 megatons of waste daily, which is a 33% increase from the quantity in 2005. The available area of waste disposal at Dhapa has been increased to 35 ha from 21 ha in 2005. The challenge is to utilize this area effectively. Door-to-door collection of wastes now covers 65% households under the municipal area, which is an increase of the figure of 50% in 2005. Number of tricycles, bin carts, contracted labourers, containers for garbage storage, transportation vehicles, trash bins have all increased from their respective figures in 2005. Wired fencing has been put up around the landfill site at Dhapa for a stretch of 6.5 km. The compost plant under KMC, which had been closed due to legal issues, has been reopened. A new biocomposting plant has also been constructed at Dhapa, covering an area of 1.5 acres [18]. In this regard the use of a decentralized biogas plant is the most viable and cost-effective route. It has been estimated that approximately 6 l of biogas can be produced daily in a 20 l reactor being fed with sufficient amount of substrates, which is approximately 10–12 kg of food waste mixed with sufficient quantity of water as required for hydrolysis. Each family has been estimated to use approximately 350–400 g of LPG per day which can be totally replaced by the quantity of biogas produced.

References

1. Joshi, N., Khatri, S., Tomar, R. K., & Jain, S.K. Greenhouse gas emissions in present and proposed municipal solid waste management plans and technologies in India: A comparative analysis of CO_2 equivalent emissions from centralized and decentralized municipal solid waste management.
2. Kumar, P., & Kaushal, R. K. Avenues of collection and disposal of municipal solid wastes management in India—A review.
3. Das, A., Sanyal, M., Roy, P. K., Majumder, A., Biswas, A. K., & Mazumdar, A. (2010). Municipal solid waste management in Kolkata Metropolitan Areas—A case study. *Environmental Science: An Indian Journal, 6*(5), 247–256.
4. Das, S., & Bhattacharyya, B. K. (2013). Municipal solid waste characteristics and management in Kolkata, India. In *The 19th International Conference on Industrial Engineering and Engineering Management* (pp. 1399–1409). Berlin, Heidelberg: Springer.
5. De, S., & Debnath, B. (2016). Prevalence of health hazards associated with solid waste disposal—A case study of Kolkata, India. *Procedia Environmental Sciences, 35,* 201–208.
6. Enerdata Independent Research.
7. Bhol, J., Sahoo, B. B. & Mishra, C. K. (2011). Biogas digesters in India: A Review.
8. Karve, A. D. Biogas in India——Samuchit Enviro Tech.
9. Pohekar, S. D., Kumar, D., & Ramachandran, M. (2005). Dissemination of cooking energy alternatives in India—A review. *Renewable and Sustainable Energy Reviews.*
10. Reddy, B. S. (2003). Overcoming the energy efficiency gap in India's household sector. *Energy Policy, 31,* 1117–1127.
11. NSSO. (1998). Results of the National Sample Survey Organization for the household sector. New Delhi: National Sample Survey Organization.

12. MNES. (2003). Annual report. New Delhi. Ministry of non-conventional energy sources.
13. WII Winrock International India. (2003). Available from: http://www.renewingindia.org.
14. Satpathy, P., Steinigeweg, S., Cypionka, H., & Engelen, B. (2016). Different substrates and starter inocula govern microbial community structures in biogas reactors. *Environmental Technology, 37*(11), 1441–1450. https://doi.org/10.1080/09593330.2015.1118559.
15. Satpathy, P., Biernacki, P., Uhlenhut, F., Cypionka, H., & Steinigeweg, S. (2016): Modeling anaerobic digestion in a biogas reactor: ADM1 model development with lactate as an intermediate (Part I). *Journal of Environmental Science and Health, Part A*. https://doi.org/10.1080/10934529.2016.1212558.
16. Solid Waste Management Services, Kolkata Municipal Corporation. https://www.kmcgov.in/KMCPortal/jsp/Solid_Waste_Services.html.
17. Farouk, H., Lang, A., Bakheet, K., Salah, M., Elobaid, M., & Abdelazim, O. (2017). Design of a household anaerobic digester for rural areas in Sudan. *International Journal of Energy Applications and Technologies, 4*, 53–63.
18. Kolkata Municipal Corporation. https://www.kmcgov.in/KMCPortal/outside_jsp/KMC_starts_portable_Compactor_15_01_.

An Experimental Study on the Generation of Biogas Using Food Waste and Water Hyacinth

Soundararaj Manju Soniya and J. Senophiyah-Mary

Abstract Globally, the energy requirements increase every day, which pose the risk of energy crisis in the future, as majority of the current need is met by crude oil. Further, environment pollutions caused by waste materials are a major concern. This study was carried to find the suitability of food waste and water hyacinth for the production of biogas at mesophilic temperature. In India food wastes are produced in large quantities. They are disposed off in open dump yard which create many health problems and also one of the main source of methane which is GHG. Another serious issue prevailing in both developed and developing countries is the disposal of the perennial aquatic plant called water hyacinth. These plants are normally disposed in open dumps which also create many serious environmental problems. A digester has been designed to treat both the food waste and water hyacinth together for the generation of biogas. The experiment was carried out for a period of 40 days. Three experimental setups were prepared with different ratios of food waste and water hyacinth with a total volume of 14 L. Parameters such as total solids, total volatile solids and pH were analyzed every day. It was found that the generation of biogas increased with increased concentration of water hyacinth. This proved that the management of water hyacinth could be increased by biogas generation which could convert waste to fuel.

Keywords Water hyacinth · Food waste · Anaerobic digestion · Mesophilic temperature · International society of waste management · Air and water

S. Manju Soniya (✉) · J. Senophiyah-Mary
Government College of Technology, Coimbatore, Tamil Nadu, India
e-mail: manju.s.soniya@gmail.com

J. Senophiyah-Mary
e-mail: senophiyah.mary8@gmail.com

J. Senophiyah-Mary
Water Institute, Karunya Institute of Technology, Coimbatore, India

© Springer Nature Singapore Pte Ltd. 2020
S. K. Ghosh (ed.), *Energy Recovery Processes from Wastes*,
https://doi.org/10.1007/978-981-32-9228-4_15

179

1 Introduction

A lot of greenhouse gases (GHG) gets emitted during the production of energy from fossil fuel which itself gets depleted. There is a huge need in the development of technology for the production of alternate fuel which could be environmentally friendly and sustainable. Solid Waste Management in Tamil Nadu is quite challenging for the urban Local Bodies (ULB) though various legislations have been passed. The improper disposal of waste water into the water bodies have increased the growth of water hyacinth which increases BOD thereby reduces the dissolved oxygen concentration which is harmful to the aquatic creatures. The depletion of fossil fuels and the increase of waste generation have paved a way for the biogas production [1].

Food waste which has a lot of proteins, carbohydrates, fats, moisture content, fiber, and vitamins could not be treated by incineration, land fillings, aerobic decomposition or could be converted into fodder for animals because none of them are environmentally sound management [2]. Utilization of food waste for the production of energy is an essential substitute for fossil energy resources [3]. As food waste has lot of organic substances they can be used for the production of biogas or biodiesel [1]. The food waste also has a few amounts of toxic substances which could be problematic in the production of biogas. Prior pre-treatment would be necessary for the production of biogas. For fermentative production of hydrogen production food waste is essential, because of its rich carbohydrate and easily hydrolysable [4]. Even water hyacinth can able to remove pollutant in water [5], basic dyes [6] and trace elements [7], negative effect can also form due to its wide growth rate. Under favorable condition, rapid production of water hyacinth even up to 60 cm/month cause low oxygen level which is dangerous to aquatic organisms and cause serious problems in navigations, irrigation and power generations [8]. Addition of waste paper into the mixture of cow dung and water hyacinth improves biogas production [9].

Widely used easy method for organic treatment is anaerobic digestion [10]. Biogas is the renewable energy that can be even produced from household and various types of waste water [11], cattle waste [12], fish industry [13] and from market and food industry waste [14]. Production of biogas requires bioreactor system which reduce greenhouse effects, decrease unpleasant odor, and even heat and power can be produced [15]. Even fruit and vegetable waste increases biogas production [16]. Methanogenic activity inhibited by high nitrogen and heavy metals and long chain fatty [17]. Investigations are carried out for rapid production of biogas [18]. Even reactor can be designed for eliminating methanogenic activity for improvement of biogas production [19]. For higher volumetric loading within smaller space and efficient gas production, high-solids anaerobic digester was also investigated [20].

2   Materials and Methods

2.1   Feed Stocks

2.1.1   Collection of Food Waste

Food wastes collected from college hostel were grinded for size reduction and mixed thoroughly by hand to make homogeneous mixture. The laboratory studies are being conducted on food waste to evaluate the chemical properties such as pH, Total solids and Volatile solids and the corresponding values are observed as 6.67, 9.3 and 94.9% respectively.

2.1.2   Collection of Water Hyacinth

Water hyacinth was collected from Srinath lake, which is located in Coimbatore south taluk, Coimbatore. Whole plant was cleaned with fresh water and sundried. The dried water hyacinth was grinded to powder. The laboratory studies are being conducted on water hyacinth to evaluate the chemical properties such as pH, Total solids and Volatile solids and the corresponding values are observed as 6.4, 16.89 and 82.95% respectively.

2.2   Inoculum

Cow dung was mixed with demineralised water in the ratio of 1:1 to use as inoculums for the culture of microorganisms so that when fresh food waste is added to it biogas, production is enhanced. The laboratory studies are being conducted on inoculum to evaluate the chemical properties such as pH, Total solids and Volatile solids and the corresponding values are observed as 6.66, 15.33 and 51.84% respectively.

3   Experimental Set up

Anaerobic reactor are designed for the biogas production even domestic waste water can be treated [21]. The experimental setup consists of three Digesters each containing the following three units. (1) A digester (2) A gas collecting unit (3) A measuring unit as shown in Fig. 1. A plastic can of 20 L capacity is used as a digester which consists of one inlet to add the feedstock and two outlets, one at top for transferring the produced gas to the gas collecting unit and another at bottom for the removal of digested slurry. Gas collecting unit is a plastic can, with brine solution inside which is used to collect the gas from the digester through a gas pipe from the digester.

Fig. 1 Anaerobic digester
fed with inoculums

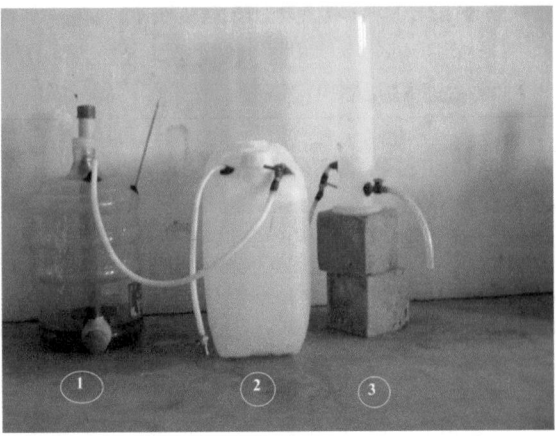

The gas from the digester compresses the brine solution which is then transferred to the measuring jar through an outlet provided at bottom of gas collecting unit. The Anaerobic Digester fed with inoculum is shown in Fig. 1.

3.1 Experimental Procedure

Three experimental Digesters were prepared with three different ratios of food waste and water hyacinth using a 20 L plastic can. Even from tubular digester [22] and advanced reactor [23], biogas can be produced. The entire content for each digester was initially mixed manually. The can was filled with a total capacity of 15 L in all the three Digesters. Table 1 shows the composition of each Digester.

3.2 Operation of the System

The digester was first loaded with inoculum and then with the feedstock through the inlet. The digester was then mixed manually in order to make contact between the biomass and the organic matter present in the digester. The digestion process was initially carried out at room temperature as mesophilic temperature. Figure 1 shows the anaerobic digester at first day of feeding.

Microorganisms the absence of oxygen break down the biodegradable material through hydrolysis, acidogenesis, acetogenesis and methanogenesis processes [24] in the digester to produce gases, which gets collected in the gas collecting unit. Since the gas collecting unit and the measuring jar are interconnected, the pressure from the produced gas compresses equal volume of brine solution to the measuring jar indicating the volume of gas produced.

Table 1 Composition of digesters

Materials	Quantity (kg)	Water proportion
Composition of digester 1		
Inoculum	1	1:1
Food waste	7	1:0.8
Water hyacinth	–	–
Composition of digester 2		
Inoculum	1	1:1
Food waste	6.6	1:1
Water hyacinth	0.4	1:1
Composition of digester 3		
Inoculum	1	1:1
Food waste	6.4	1:1
Water hyacinth	0.8	1:1

4   Results

4.1   Total Solids

The total solids for the samples were determined experimentally from day 4 to day 40 using standard methods from the initial day of feeding. At day 4, total solids in digester 1 is 14.128%, digester 2 is 18.8% and in digester 3 is 21.33% which is gradually decreased in all digester to 6.24, 6.39 and 7.21% respectively. The reduction of total solids content noticed in all the three Digesters are graphically represented of reduction in Fig. 2.

Fig. 2 Reduction of total solids

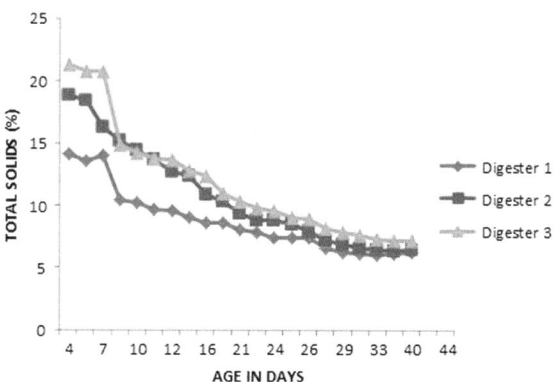

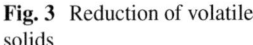

Fig. 3 Reduction of volatile
solids

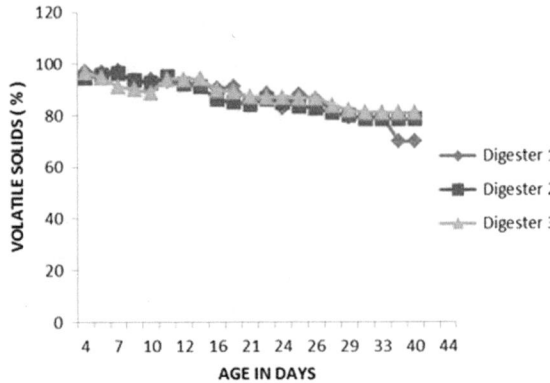

4.2   Volatile Solids

The volatile solids content for the samples were also determined experimentally
through standard laboratory methods. Initially Volatile solids found to be slight
increase in digester 1 and 2 but started to gradually decrease afterward. Gradual
decrease of volatile solids digester 3 is noticed. Volatile solids in digester 1 reduced
from 96.91 to 69.84% and in digester 2 from 94.55 to 78.52% and for digester 3 from
96.87 to 80.2%. The graphical representation of the reduction in volatile solids with
respect to time is shown in Fig. 3.

4.3   Temperature

The most influencing parameter for anaerobic digester is temperature. It influences both enzymes activity and biogas production. Usually anaerobic bacteria grow
between 10° and 60°, whose performance increasing with the increasing in temperature [24]. Initially room temperature was found in all the digester which is gradually
increased day by day. Temperature of the sludge during digestion was found to be
35 ± 2 °C for mesophilic and 50 ± 2 °C for thermophilic condition to occur. Temperature affects the chemical equilibrium by producing new substance through chemical
reaction leads to decrease of pH level [18]. The inside temperature of the three
digesters were noted and the readings are given in the graphical form with respect
to time in Fig. 4. It also founds that biogas production at thermophilic condition is
higher than mesophilic condition [25].

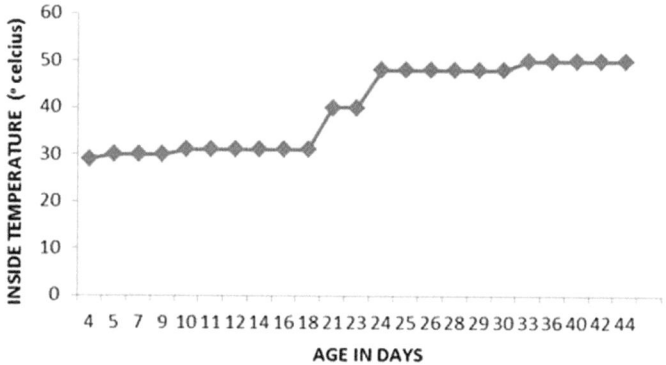

Fig. 4 Temperature variations

4.4 pH

Another important factor affecting anaerobic digestion is pH. Different pH range required by anaerobic bacteria for their growth. It was found to be 4–8.5 ranges required for fermentative bacteria. The favorable range for methanogenesis growth was found to be 6.5–7.2 [24]. The pH value of the feedstock was noted from the initial day of loading till the end of process. The pH value of the digester found to be reduced during initial stage of loading hence sodium hydroxide solution was added to all the three Digesters into neutralize the pH between 6 and 8 to attain a favorable medium for the growth of anaerobic bacteria [4].

4.5 Gas Production

The gas production is the important parameter in the anaerobic digestion process. The gas production measured every day and the gas burnt with blue flame indicated the presence of methane gas. To obtain efficient biogas production, different techniques for mixing, monitoring and controlling is necessary [1]. The variation of gas collection with time period is presented graphically in Fig. 5.

5 Conclusion

Anaerobic digestion is a reliable technology to digest food waste and Water Hyacinth effectively under mesophilic and thermophilic condition in an economic way. The following observation was found by a laboratory scale anaerobic digester of 20 L capacity to evaluate the yield of biogas from food water and dried water hyacinth

Fig. 5 Cumulative gas produced

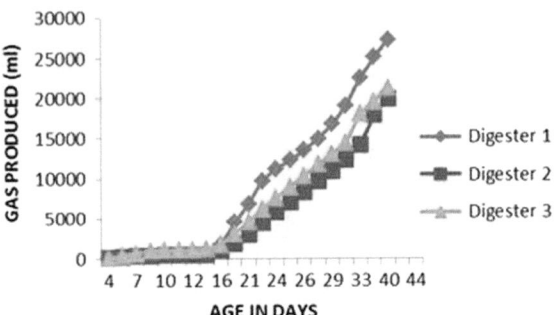

On 4th day digester 1 and 3 produces same amount of gas and digester 3 starts to produce greater biogas than digester 1, which shows 3.75% increased biogas production up to 16th day afterward the trend gets altered. For instance on 23rd day Digester 1 shows 35% increased bio gas production when compared to Digester 3. It is observed that there is a significant increase in gas production in all the ages except the initial day in Digester 3 when compare to Digester 1. Maximum observation found in 4th day and it almost 1.3 times more value that Digester 2. Initially digester 2 produces greater biogas than digester 1 afterward trends change. At 12th day digester 1 and digester 2 produces same amount of Biogas. Total solids found to be gradual decrease in all digester at day 4, total solids in digester 1 is 14.128%, digester 2 is 18.8% and in digester 3 is 21.33% which is gradually decreased to 6.24, 6.39 and 7.21% respectively on 40th day. Gradual decrease of volatile solids digester 3 is noticed. Volatile solids in digester 1 reduced from 96.91 to 69.84% and in digester 2 from 94.55 to 78.52% and for digester 3 from 96.87 to 80.2%. Food waste and Water Hyacinth was found to be desirable feedstock for anaerobic digestion. During the process of biodegradation, developed free ammonia and ammonium may create toxics to microbes and affect the performance. Hence approach for ammonia removal can be studied for future.

References

1. Weiland, P. (2010). Biogas production: Current state and perspectives. *Applied Microbiology and Biotechnology, 85,* 849–860. https://doi.org/10.1007/s00253-009-2246-7.
2. Chu, C.-F., Li, Y.-Y., Xu, K.-Q., Ebie, Y., Inamori, Y., & Kong, H.-N. (2008). A pH- and temperature-phased two-stage process for hydrogen and methane production from food waste. *International Journal of Hydrogen Energy, 33,* 4739–4746. https://doi.org/10.1016/j.ijhydene.2008.06.060.
3. Klocke, M., Nettmann, E., Bergmann, I., Mundt, K., Souidi, K., Mumme, J., et al. (2008). Characterization of the methanogenic Archaea within two-phase biogas reactor systems operated with plant biomass. *Systematic and Applied Microbiology, 31,* 190–205. https://doi.org/10.1016/j.syapm.2008.02.003.

4. Kim, S. (2004). Feasibility of biohydrogen production by anaerobic co-digestion of food waste and sewage sludge. *International Journal of Hydrogen Energy, 29*, 1607–1616. https://doi.org/10.1016/j.ijhydene.2004.02.018.
5. Muramoto, S., & Oki, Y. (1983). Removal of some heavy metals from polluted water by water hyacinth (*Eichhornia crassipes*). *Bulletin of Environmental Contamination and Toxicology, 30*, 170–177. https://doi.org/10.1007/BF01610117.
6. Low, K. S., Lee, C. K., & Tan, K. K., n.d. Biosorption of basic dyes by water hyacinth.
7. Zhu, Y. L., Zayed, A. M., Qian, J.-H., de Souza, M., & Terry, N. (1999). Phytoaccumulation of trace elements by wetland plants: II. Water hyacinth. *Journal of Environment Quality, 28*, 339. https://doi.org/10.2134/jeq1999.00472425002800010042x.
8. Malik, A. (2007). Environmental challenge vis a vis opportunity: The case of water hyacinth. *Environment International, 33*, 122–138. https://doi.org/10.1016/j.envint.2006.08.004.
9. Yusuf, M. O. L., & Ify, N. L. (2011). The effect of waste paper on the kinetics of biogas yield from the co-digestion of cow dung and water hyacinth. *Biomass and Bioenergy, 35*, 1345–1351. https://doi.org/10.1016/j.biombioe.2010.12.033.
10. Zhang, R., Elmashad, H., Hartman, K., Wang, F., Liu, G., Choate, C., et al. (2007). Characterization of food waste as feedstock for anaerobic digestion. *Bioresource Technology, 98*, 929–935. https://doi.org/10.1016/j.biortech.2006.02.039.
11. Lettinga, G., & Hulshoff Pol, L. W. (1991). UASB-process design for various types of wastewaters. *Water Science and Technology, 24*, 87–107. https://doi.org/10.2166/wst.1991.0220.
12. El-Mashad, H. M., & Zhang, R. (2010). Biogas production from co-digestion of dairy manure and food waste. *Bioresource Technology, 101*, 4021–4028. https://doi.org/10.1016/j.biortech.2010.01.027.
13. Kafle, G. K., Kim, S. H., & Sung, K. I. (2013). Ensiling of fish industry waste for biogas production: A lab scale evaluation of biochemical methane potential (BMP) and kinetics. *Bioresource Technology, 127*, 326–336. https://doi.org/10.1016/j.biortech.2012.09.032.
14. Brown, D., & Li, Y. (2013). Solid state anaerobic co-digestion of yard waste and food waste for biogas production. *Bioresource Technology, 127*, 275–280. https://doi.org/10.1016/j.biortech.2012.09.081.
15. Widodo, T. W., & Asari, A. n.d. Design and development of biogas reactor for farmer group scale. *Indonesian Journal of Agriculture.*
16. Shen, F., Yuan, H., Pang, Y., Chen, S., Zhu, B., Zou, D., et al. (2013). Performances of anaerobic co-digestion of fruit & vegetable waste (FVW) and food waste (FW): Single-phase vs. two-phase. *Bioresource Technology, 144*, 80–85. https://doi.org/10.1016/j.biortech.2013.06.099.
17. Haider, M. R., Zeshan, Yousaf, S., Malik, R. N., & Visvanathan, C. (2015). Effect of mixing ratio of food waste and rice husk co-digestion and substrate to inoculum ratio on biogas production. *Bioresource Technology 190*, 451–457. https://doi.org/10.1016/j.biortech.2015.02.105.
18. De la Rubia, M. Á., Walker, M., Heaven, S., Banks, C. J., & Borja, R. (2010). Preliminary trials of in situ ammonia stripping from source segregated domestic food waste digestate using biogas: Effect of temperature and flow rate. *Bioresource Technology, 101*, 9486–9492. https://doi.org/10.1016/j.biortech.2010.07.096.
19. Zhu, H., Parker, W., Conidi, D., Basnar, R., & Seto, P. (2011). Eliminating methanogenic activity in hydrogen reactor to improve biogas production in a two-stage anaerobic digestion process co-digesting municipal food waste and sewage sludge. *Bioresource Technology, 102*, 7086–7092. https://doi.org/10.1016/j.biortech.2011.04.047.
20. Guendouz, J., Buffière, P., Cacho, J., Carrère, M., & Delgenes, J.-P. (2010). Dry anaerobic digestion in batch mode: Design and operation of a laboratory-scale, completely mixed reactor. *Waste Management, 30*, 1768–1771. https://doi.org/10.1016/j.wasman.2009.12.024.
21. Van Haandel, A., Kato, M. T., Cavalcanti, P. F. F., & Florencio, L. (2006). Anaerobic reactor design concepts for the treatment of domestic wastewater. *Reviews in Environmental Science and Biotechnology, 5*, 21–38. https://doi.org/10.1007/s11157-005-4888-y.
22. Bouallagui, H. (2003). Mesophilic biogas production from fruit and vegetable waste in a tubular digester. *Bioresource Technology, 86*, 85–89. https://doi.org/10.1016/S0960-8524(02)00097-4.
23. Lettinga, G., & Pol, L. H. (1986). Advanced reactor design, operation and economy. *Water Science and Technology, 18*, 99–108. https://doi.org/10.2166/wst.1986.0166.

24. Zhang, C., Su, H., Baeyens, J., & Tan, T. (2014). Reviewing the anaerobic digestion of food waste for biogas production. *Renewable and Sustainable Energy Reviews, 38,* 383–392. https://doi.org/10.1016/j.rser.2014.05.038.

25. Vindis, P., Mursec, B., Janzekovic, M., & Cus, F. (2009). The impact of mesophilic and thermophilic anaerobic digestion on biogas production. *Journal of Achievements in Materials and Manufacturing Engineering, 36,* 192–198. 10.1.1.466.555.

Bioethanol Production from Natural Plant Substrates of Terrestrial Source

R. JayaMadhuri, M. Saraswathi, K. Gowthami, M. Sujatha and T. Uma

Abstract Petroleum-based fuels are non-renewable resources, causing severe air pollution to the environment. Hence, biofuels are given much importance as they are sustainable, produced from biological forms, cost-effective and eco-friendly. Bio-ethanol is a volatile and flammable liquid produced through the microbial fermentation process. Bio-ethanol is a microbiological way of converting simple sugar into ethanol and carbon dioxide. This work focused on the economically viable source used for the production of bio-ethanol from agrowaste leaf extract used as a substrate like agave leaves, neem leaves and cassava fermented *by Aspergillus niger, Trichoderma viride* and *Saccharomyces cerevisiae*. The fermented product was purified by distillation process of ethanol which was determined by alcoholmeter, and the bio-ethanol was characterized by FTIR analysis.

Keywords Bio-ethanol · Non-renewable · Biofuels · Petroleum · Alcoholmeter

1 Introduction

Petroleum-based fuels are non-renewable resources, causing severe air pollution to be the environment. Hence, biofuels are given much importance as they are sustainable, produced from biological forms, cost-effective and eco-friendly. Among the biofuels, bio-ethanol seems like a good alternative to fossils fuel and termed as a renewable energy source since its source crops can be grown continuously, leading to lower emission of greenhouse gases and no threat to our ecosystem [1]. Bio-ethanol is a volatile and flammable liquid produced through the microbial fermentation process. It has a molecular formula C_6H_5OH. It can be sufficiently used as an energy source for transportation [2]. The most common blend is 10% ethanol and 90% petrol (E), and vehicle engines require no modification to run on ethanol [3]. Ethanol is liquid biofuel produced from sugar-rich biomass, and ethanol can be blended upto 20% with desiel or petrol. Moreover, ethanol fuel is a substitute to gasoline and is produced

R. JayaMadhuri (✉) · M. Saraswathi · K. Gowthami · M. Sujatha · T. Uma
Department of Applied Microbiology, Sri Padmavathi MahilaVisvavidhyalayam,
Tirupati 517501, India
e-mail: drjayaravuri@gmail.com

© Springer Nature Singapore Pte Ltd. 2020
S. K. Ghosh (ed.), *Energy Recovery Processes from Wastes*,
https://doi.org/10.1007/978-981-32-9228-4_16

from the translation of carbon-based feedstocks. Ethanol is a correctly used gasoline for vehicles.

Most vehicles run in a mixture of gas ethanol; this mixture is common at most gas stations across the country [4]. Bio-ethanol is a microbiological way of converting simple sugar into ethanol and carbon dioxide. The most important resources of sugar are necessary to manufacture ethanol coming from fuel or energy crops. It is majorly produced in batch fermentation with fungi strain such as *Aspergillus and Saccharomyces cerevisiae* that cannot tolerate high concentration ethanol [5]. In the present-day research on the run edible plant source, neem tree leaves are used in bio-ethanol production using baker's yeast as a fermenting agent prior to the acid hydrolyzed substrate. In that view of above, the following study was aimed to investigate the potential of neem tree leaves (*Azadirachta indica*) in bio-ethanol production using baker's yeast *Saccharomyces cerevisiae* as the fermenting agent. The hydrolyzed and filtered extracts were fermented products purified by distillation process, and the presence of ethanol was determined by alcoholmeter method [6]. Hence, in our present study, agave leaves were used as raw material for the production of bio-ethanol using Saccharomyces cerevisiae and *Trichoderma viride*. Biofuels such as bio-ethanol can be produced by many different types of substrates among these cassavas (Manihot enculenta Crantz), a plant with high starch content, is considered a cheap, edible, abundant and promising resource for the production of fermentable glucose syrups and dextrins. Moreover, it is easily produced in tropical and subtropical zones. This research work, therefore, examines the production of bio-ethanol from cassava peels.

Immobilization technique is a method used to entrap microbial cells/enzymes and keep them stable for prolonged activity. In the present study, sodium alginate method has been used to immobilize microbial species used in the experiment. Experimental data are compared with normal cultures in terms of yield and production period. Further, scale-up studies were much important in fermentation economics. Finally, in the research work, it has been aimed to produce the economically important biofuel that is bio-ethanol from the non-expensive natural solid waste substrates, namely neem, agave and cassava and compare their yield.

1.1 Materials and Methods

1.1.1 Sample Collection

Neem tree leaves were collected from SPMVV campus, agave leaves were procured from TTD nursery in the campus, and cassava sample was obtained from local area Indira Priyadarshini market, Tirupati (Fig. 1).

Fig. 1 Collected samples for the production of bio-ethanol

1.1.2 Sample Pretreatment and Preparation

Neem

The neem tree leaves were collected from SPMVV ground. Neem leaves were collected in polythene bags, and neem leaves were air-dried and grounded into powder followed by sieving process. Three grams of powdered neem leaves was taken into conical flasks and pretreated using H_2SO_4. The flasks were sterilized at 121 °C for 15 min, and then, Saccharomyces inoculum was inoculated into pretreated solid substrate.

Agave

The substrate (*Aloe barbadensis*) was collected from agricultural land. Boiled agave was crushed, and it was used for ethanol production. The pretreated substrate was done by cooking agave leaves (100 g) in a pressure cooker in 100 ml water containing 0.5% potassium meta bisulphate [7].

Cassava

Hundred grams (100 g) of cassava peels was collected from sweet potatoes. The cassava peel was air-dried and grounded into powder followed by sieving. The potato dextrose agar media slants were prepared and then inoculated with *Aspergillus Niger*. After 4 days, freshly harvested cells were obtained for fermentation process.

Fermentation process—Neem

For the production of bio-ethanol from neem leaves, the substrate is added with Saccharomyces cerevisiae. Fermentation medium was incubated aerobically at 37 °C for five days. Then, the estimation of reducing sugar and amount of alcohol produced was performed.

Agave

The hydrolysed and filtered extract was fermented using *Saccharomyces cerevisiae* for 7 days of incubation at room temperature under anaerobic condition carried out in round bottom flask at 80 °C. After that, percentage of ethanol was estimated.

Enzyme production and assay

Aspergillus niger and *Trichoderma viride* were inoculated in potato dextrose broth and incubated at room temperature for 7 days. After incubation, fermentation media was filtered by Whatman No. 1 filter paper and centrifuged, and the supernatant was collected for enzyme assay—amylase and cellulose.

Enzymatic hydrolysis

Twenty grams of agave leaves was crushed and dissolved in 200 ml of distilled water. The content was boiled and filtered through Whatman No. 1 filter paper. The extract was sterilized, and then, enzyme source was added to the extract and incubated at 37 °C for 24 h for hydrolysis process.

Cassava

Twenty-five grams of cassava peel powder was taken and then added 100 ml of distilled water into that and autoclaved. It was cooled and then added the inoculums (*Aspergillus niger*) and incubated for 7 days. After incubation, *Saccharomyces cerevisiae* was added to the fermentation medium and incubated for 7 days, and the estimation of alcohol was done.

Estimation of reducing sugars in Neem and Agave by dinitrosalicylic method

Different volumes of glucose solution from 0.2 to 1.0 ml were pipetted out into various test tubes, and the volume was made to 2 ml with water. To each tube, 2 ml of DNS reagent was added and kept in boiling water bath 10 min after the tubes were cooled and diluted with 10 ml of water. The orange red colour flourished was measured at 520 nm by using a reagent as blank. The colour intensity indicates the biological sample concentration.

Determination of total and residual sugars

The total sugar content of agave leaves was determined by phenol sulphuric acid method [8]. The reducing sugar percentage was calculated by subtracting the residual sugar [9] from total sugar.

Phenol sulphuric acid method

0.2, 0.4, 0.8 and 1 ml of working standard were pipetted out into a series of test tubes. Similarly, 0.1 and 0.2 ml of the sample solution were pipetted out in two separate test tubes and the volume in each tube was made to 1 ml with distilled water. Set a blank with 1 ml of water. One ml of phenol was added to each tube. Five ml of 96% sulphuric acid was added to each of the tube and shaken well. After 10 min, the content in the tubes are placed in a water bath. The colour intensity was read at 490 nm. The amount of total carbohydrates present in the sample solution was calculated using the standard graph.

Estimation of alcohol by potassium dichromate method [10]

Aliquots of given sample ranging from 0.2 to 1.0 ml were taken. 1 ml of $K_2Cr_2O_7$ solutions was added. The solution was mixed and cooled by pouring ice cold water, and 4 ml of conc. H_2SO_4 was added along the walls of the test tube collected. This is highly endothermic reaction after the tubes were cooled, and OD values were taken at 660 nm.

Recovery of bio-ethanol [11]

Bio-ethanol formed in the fermentation broth was collected using a rotary evaporator available at DST Curie Centre, SPMVV, Tirupati. Another flask was fixed to the other end of distillation column to collect the distillate at the standard temperature for ethanol production. The distillate collected was measured using a measuring cylinder, and the quantity of ethanol produced is expressed in millilitres/percentage.

Entrapment method

0.1 g of NaCl in 100 ml of distilled water was added. Added 3 g of sodium alginate to four percentage of $CaCl_2$ was prepared in 100 ml of distilled water. Cool this to 15–20 °C. Culture alginate slurry was dissolved in small quantity of distilled water to get cell paste. After 1 h of incubation, sodium alginate solution was taken and culture was slowly added with constant stirring to avoid bubble formation. This makes the cell to gel entrapped in mattress of alginate slurry. The slurry was taken in aboard tip Pasteur pipette syringe whose needle was removed and checked before for proper control and droplet formation. The slurry taken in syringe or Pasteur pipette for slowly dropping in the form of droplet into $CaCl_2$ gives stability to the sodium alginate. Gel drops in which the cells were entrapped.

2 Results

Total reducing sugar estimation was carried out by phenol sulphuric acid method using untreated agave leaves. The maximum total reducing sugar percentage of neem

Table 1 Amount of reducing sugars estimated by DNS method

S. No.	Concentration in mg/ml	OD values
Standard	0.2	0.31
	0.1	0.11
	0.6	0.52
	0.8	0.63
	1.0	0.75
Neem	0.63	0.55
Agave	1.05	0.79

Fig. 2 Colour intensity of reducing sugars in neem leaves

leaves was given in Table 1 and Fig. 2. As summarized in Table 4, Agave leaves contain more sugar concentration than neem.

Alcohol estimation: The concentration of ethanol present in the three substrates was found to be higher in cassava followed by neem and minimum in agave (Tables 2 and 3).

Bio-ethanol produced by immobilized Saccharomyces and Trichoderma sp in neem, agave and cassava plant substrates

The results obtained in sodium alginate entrapment method were represented in Table 4. Significant increase in yield and reduction in incubation period are observed in all three samples tested.

Table 2 Alcohol estimation in the samples

S. No.	Concentration in %	OD values
Standard 1	20	0.47
2	40	0.54
3	60	0.65
4	80	0.76
5	100	0.83
Neem	38.5	0.52
Agave	18.7	0.44
Cassava	60	0.65

Table 3 Estimation of total sugars by phenol sulphuric acid method in agave

S. No.	Concentration in mg/ml	OD values
Standard	0.2	0.25
	0.4	0.33
	0.6	0.48
	0.8	0.54
	1.0	0.67
Cassava	0.45	0.39

Table 4 Effect of immobilization on bio-ethanol production

S. No.	Bio-ethanol production without entrapment	Bio-ethanol production with entrapment
Neem	38.5 (5 days)	58.5 (2 days)
Agave	18.7 (7 days)	33.6 (3 days)
Cassava	60 (7 days)	77.8 (3 days)

Scale-up studies

Initially, the experiments were conducted using 100 ml fermentation; medium volume of the medium has been increased gradually to 250 ml and 500 ml, and the yield was indicated in Table 5.

TIR analysis

The samples (neem, agave and cassava) of bio-ethanol produced were analysed using MB 3000 FTIR spectroscopy machine to determine the vibrational frequencies of the bio-ethanol produced (Figs. 2, 3, 4, 5 and 6). The highest yield bio-ethanol obtained that is using acid hydrolysis was subjected to FTIR spectroscopy analysis, and the samples have shown O–H stretching bond peak at 3597–3650. O–H stretching is due to the presence of phenol group.

Table 5 Quantitative estimation of ethanol in scale-up of fermentation medium

S. No.	Composition of medium (ml)	Yield of ethanol in %
1. Neem	100	38.5
	250	42.3
	500	57.4
2. Agave	100	18.7
	250	23.5
	500	29.6
3. Cassava	100	60
	250	71.4
	500	68.5

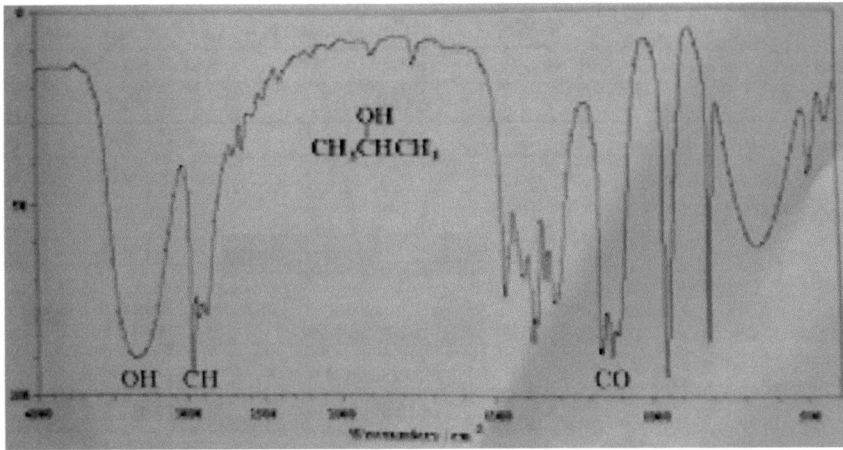

Fig. 3 FTIR spectra of standard bio-ethanol

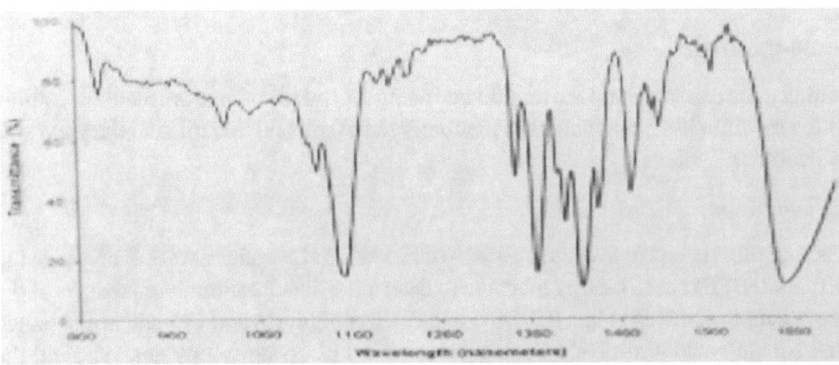

Fig. 4 FTIR spectrum of bio-ethanol produced from neem

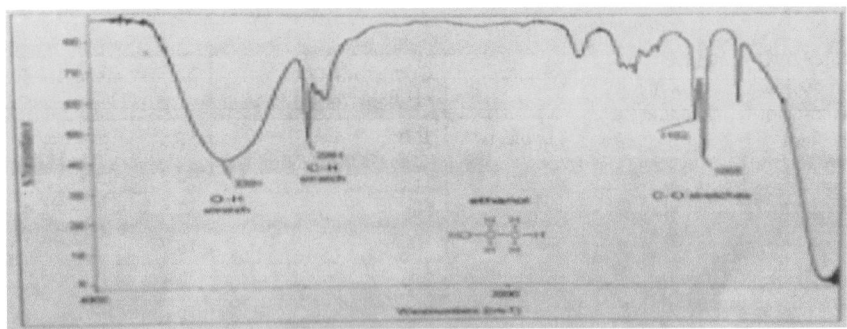

Fig. 5 FTIR spectrum of bio-ethanol produced from agave

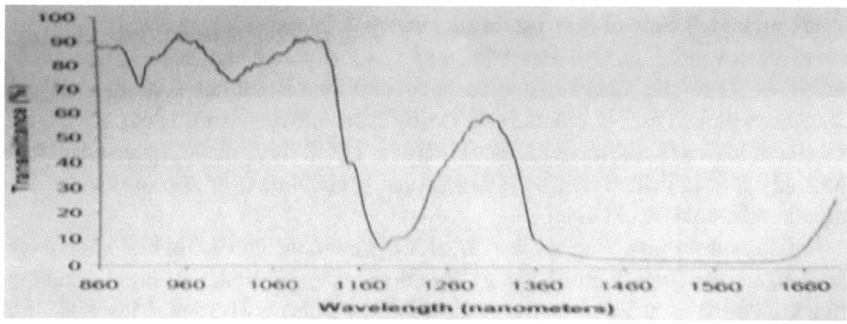

Fig. 6 FTIR spectrum of bio-ethanol produced from cassava

3 Discussion

Bio-ethanol is a principle fuel that can be used as a petrol substitute for the vehicle. It is a renewable energy source produced mainly by the sugar fermentation process, although it can also be manufactured by the chemical process of reacting ethylene with steam. These crops include maize, cassava products and wheat crops, waste straw, rice husk, millet husk, guinea corn husk and sorghum plant. Ethanol is high-octane fuel and has replaced lead as an octane enhancer in petrol. By blending ethanol with gasoline, we can also oxygenate the fuel mixture so it burns more completely and reduces pollution emission. Ethanol has been produced in batch fermentation with fungi strains such as *Aspergillus niger, Mucor mucedo* and *Saccharomyces cerevisiae* that cannot tolerate high concentration of ethanol biofuels produced by many different types of substrates.

In recent years, the production of ethanol using fermentation on a large scale has attained considerable interest. The economic feasibility however always has been focused towards high yield of ethanol that is full use of raw material associated with high productivities so as to reduce the cost of production. The crude enzymes like amylase and cellulase were used for the enzymatic hydrolysis of biomass. In the hydrolysis process, few extracts of biomass and whole biomass were treated with the crude enzymes.

Purification of the crude enzymes and optimization parameters may give a better result for the degradation of starch of hemicellulose and cellulose present in biomass. The maximum ethanol percentages of amylase-treated (*Saccharomyces cerevisiae*) agave leaves and cellulose-treated agave leaves were found, respectively. This is due to the fact that an organism has several advantages over yeast like higher rates of glucose uptake to ethanol production, higher ethanol yields and ethanol tolerance.

Bio-ethanol was produced from agave leaves using an enzymatic hydrolysis process and enzyme hydrolysis (amylase and cellulose) followed by fermentation. The results showed that the cassava and neem leaves yield maximum, whereas the minimum ethanol is recorded with agave leaves containing *Saccharomyces cerevisiae*.

Production of ethanol from lignocellulose material has received extensive interest due to their availability, abundance and relativity low cost. Neem leaves (*Azadirachta indica*) is, therefore, abundant and sustainable biomass and non-food material that could be exploited for bio-ethanol production especially in the northern part of Nigeria. Neem leaves (*Azadirachta indica*) however could serve this purpose since from the study it is indicated that by either proper pretreatment or appropriate method bio-ethanol could be obtained.

Entrapment and scale-up studies revealed a prominent increase in bio-ethanol production, and production time has also been drastically reduced, thus highlighting the immobilization technique in fermentation studies. This may be due to the protection of the microbial from the external environment.

Entrapment samples were subjected to FTIR Spectroscopy analysis, the samples showed a strong broad peak at 3450-2850-1, therefore indicating –CH2– and –CH3 stretching vibrations and well-resolved peak around $3416 \, cm^{-1}$ can be assigned to an alcoholic –OH– vibrations. The values are in agreement with the standard value and previous reports. According to Spectra (2014) free O–H stretching normally occurs at $3550–3200 \, cm^{-1}$, while C–H stretch occurs at $3000–2840 \, cm^{-1}$. Therefore, the production of ethanol was successful. The results obtained from the present study can be concluded that H_2SO_4 acid hydrolysis gave the highest concentration of ethanol production. This indicates that baker's yeast (*Saccharomyces cerevisiae*) is a suitable fermenting agent in the production of bio-ethanol using neem leave sample.

4   Conclusion and Future Prospective

The population of a human being is increasing on the average; hence, the demand for energy source increases. It is apparent that current fuel bio-ethanol production from grain-based feedstock is not favourable as it may lead to a food shortage to the taming world populace. In order to avoid these foreseen worrisome, lignocelluloses biomass should be utilized in the production of bio-ethanol and biofuels in general. The study showed that all the three natural plant substrates, namely neem, agave and cassava samples have potential in producing the bio-ethanol in an economical and eco-friendly manner. In the future, we want to imply different immobilization materials and check potential of bio-ethanol production. In addition, we want to check the nanoparticles as an aid in purification of bio-ethanol from the fermentation broth.

References

1. Balat, et al. (2008). Ethanol production by *Saccaromyces cerevisiae* from cassava peel hydro lysate. *Internet Journal of Microbiology*.

2. Pramanik, Lee, L. J., Tride, D. E., & Rogers, P. L. (2005). Zymomonasmobilis CP4. *Biotechnology Letters, 15*(13), 137–142.
3. Chandel, et al. (2012). Dilute acid hydrolysis agro-residues for the de polymerization of hemicellulose.
4. Prasad, et al. (2012). The removal of chromium (III) from aqueous solution. High productivity ethanol fermentations with journal of chemical technology and biotechnology, Zymomonasmobilis. *Process Biochemistry, 15*, 7–11.
5. Graeme, M. W. (2010). Bioethanol science and technology for fuel alcohol, vents publishers and as USA Application-17.
6. Garcia Reyes, B. R., & Rangel Mendez, J. R. (2009). Ethanol production by Zymomonasmobilis. Continuous culture at high glucose concentrations. Contribution of Agro waste material main Biotechnology. *Letters, 1*, 421–426. Components (hemicelluloses, cellulose and lignin) to 16.9.Rogers, P.L., K.J. Lee and D.E tribe, 1980.
7. Hill, J., Nelson, E., Tilman, D., Polasky, S., & Tiffancy, D. (2006). *Production technology* (Vol. 56, pp. 17–34.14). New Age publishers.
8. Verma, G., Nigam, P., Singh, D., & Chaudary, K. (2000). From *Aspergillusniger* JGI 24 Isolated in Bangalore (p. 980).
9. Somogyi, M. (1945). A New reagent for the determination of sugar. *Journal of Biological Chemistry, 160*, 61–98.
10. Lopez, M., Mancilla, M. N., & Mendoza Diaz, G., Pik, J., & Lawford, H. G. (2003). I Environmental. *Economic and Delhi Science USA, 103*, 11206–11210 (Production and Characterization of—amylase).
11. Stanbury. (1981). A1533-1538. 17. Leaves, B.H., P. Pang, C.R. Mackenzie, G.R. Lawford.
12. Somogyi, M. (1945). Commercialization of biofuel industry in Afica.

Bio- and Thermochemical Conversion of Poultry Litter: A Comparative Study

S. Sharma, K. Pradeepkumar, M. Shaanmaadhuran, M. J. Rajadurai, Y. Anto Anbarasu and V. Kirubakaran

Abstract More than 90% of world's population takes eggs and broiler chicken as food directly or indirectly to meet their protein demand economically. Egg and broiler industries have grown in size considerably during the last decade. Several food processing, preservation and transporting technologies have been developed, demonstrated and adopted successfully. Both heat energy and electrical energy are vital to the broiler production house. Heat is vital for the brooding period of a newly hatched chick, and heating in the winter helps optimize feed consumption and growth. Electrical energy is important for automatic feed delivery and especially for ventilation and cooling purposes. The poultry litter contains volatile matter, fixed carbon, ash and water with a calorific value of 2276 kcal/kg. Poultry litter has traditionally been land spread on soil as an amendment. Excessive spreading enriches the soil with water nutrients resulting in eutrophication of water bodies, the spread of pathogens, the production of phytotoxic substances, air pollution and emission of greenhouse gases. Therefore, identification of an alternate eco-friendly disposal route with potential financial benefits has become the need of the hour. Schemes providing energy and easy to handle fertilizer as a by-product would be the best alternate route. Hence, this paper attempts the comparative study of the bio- and thermochemical conversion of poultry litter for converting waste into biogas as well as producer gas.

Keywords Poultry litter · Biogas · Gasification · Waste-to-energy conversion

1 Introduction

India is one of the developing countries that is struggling to keep up with the energy demand for its large population. The country is producing annually about 1206.306 billion units against the demand of 1265 billion units [1]. Meeting such demands has traditionally relied on fossil fuel-based power generation plants. However, against the backdrop of rising fuel price and decreasing fuel resources, focus has shifted to

S. Sharma · K. Pradeepkumar · M. Shaanmaadhuran · M. J. Rajadurai · Y. Anto Anbarasu ·
V. Kirubakaran (✉)
The Gandhigram Rural Institute (Deemed to be University), Dindigul, India
e-mail: kirbakaran@yahoo.com

S. K. Ghosh (ed.), *Energy Recovery Processes from Wastes*,
https://doi.org/10.1007/978-981-32-9228-4_17

renewable energy resources [2]. Now biogas and gasification hold a great potential resource for energy generation. In India, according to 2018 report, chicken meat consumption was 2784 MT [3]. Huge chicken production leads to large amount of poultry waste. In India, the total removal of fuel wood from forest land is estimated to be 270 million tons annually which results in deforestation. This paper focuses on power production from poultry waste and sawdust [4, 5]. We did the sample analysis and experimental results to show the potential of power production from these wastes. Biogas is an environmentally friendly and one of the most efficient and effective option for renewable energy among various other alternative sources [6].

1. A. S. M. Mominul Hasan, an author from Bangladesh, identified that the calorific value of poultry waste is higher than the cow manure. He includes the economic feasibility analysis of biogas plant. With an existing poultry farm, he developed a biogas plant and generated 60 kWh of power from the excretion of 6000 birds. This analysis further reveals that the money from selling slurry from biogas plant and the generated electricity can recover the investment within 4–6 years of time.

2. Joseph R.V. Flora (2006) made a comparative study of using poultry litter for generating power by the different process such as anaerobic digestion, open firing, co-firing and gasification. In addition, he includes cost analysis, rate of return and payback period and discussed the best method in the terms of economic analysis. Finally, he concludes that gasification is the effective method for generating power in the terms of economic term.

3. In 'Anaerobic digestion technology in poultry and livestock waste treatment' by Suleyman Sakar, Kaan Yetilmezsoy and Emel Kocak, different anaerobic processes used in poultry and livestock treatment are discussed. The final results include a wide range of different reactor volumes from 100 to 95 mL that can be utilized to investigate the anaerobic processes in poultry manure.

4. In Illinois River watershed case study by Roger L. OlsenaRick, W. Chappellb, Jim C. Loftisc, the water quality investigations to characterize large watersheds are discussed. Further, more water quality sample collection and treatment for principal component analysis for different water samples is presented. This result reveals the different chemical and biological constituents of the waterbody.

5. In 'Anaerobic co-digestion of hog and poultry waste' by Benjamin S. Magbanua Jr., Thomas T. Adams and Phillip Johnstonc, the anaerobic batch tests were performed using hog and poultry wastes in various proportions. Treatments that received both wastes produced good yields of biogas.

6. In 'Continuous co-digestion of cattle slurry with fruit and vegetable wastes and chicken manure' by F. J. Callaghana et al., has proposed anaerobic digestion for treating both solid and liquid organic wastes. As such, the digestion of cattle slurries and agricultural wastes has been evaluated.

7. 'Anaerobic digestion of organic solid wastes—An overview of research achievements and perspectives' by J. Mata-Alvarez, S. Macé and P. Llabrés describes the technology for anaerobic digestion of organic solid wastes with many features. Topics such as essentials (kinetics, modelling, etc.), process phases (performance, two- and single-phase systems, wet and dry technologies), digestion development

(several pre-treatments), co-digestion with their other substrates and its relation to composting technology are examined in this review.

2 Process of Biogas and Gasification

2.1 Gasification

Gasification is the process of converting any kind of biomass into gaseous fuel carbon monoxide, hydrogen and carbon dioxide at a temperature 700–1100 °C with sub-stoichiometry oxygen supply/air supply. The resulting gas is termed as syngas/producer gas, which is considered as a fuel. In the initial process of gasification, generous amount of air is supplied for the preparation. The combustion is reduced to sub-stoichiometric air for effective gasification. This process involves drying zone where the moisture content is removed from the biomass (poultry waste) at 110 °C [7]. The next process is pyrolysis at 200–300 °C. In the pyrolysis zone, it separates the volatile matter from the biomass and the part of devolatilized matter that enters into the combustion chamber where enough amount of air is there for effective combusting. At the end in the reduction zone, the unburnt char combines with water vapour and volatile gas which results in gases like carbon monoxide, methane, hydrogen and carbon dioxide. Finally, the ash is formed at the bottom of the grate and the syngas is extracted from the outlet pipe and used for some combustion applications [8] (Fig. 1).

2.2 Biogas

Biogas reactor produces methane by converting liquid manure, into methane in an anaerobic condition. After processing, the remaining sludge from the organic matter can be used as a fertilizer. Biogas involves three stages, namely enzymatic hydrolysis, acid formation and gas production. At the end of organic digestion, the gaseous products that escape from the solids are methane, carbon dioxide and less amount of hydrogen sulphide [9, 10].

- In the first phase, the hydrolysis process takes place where the complex molecules are broken down into simple structure of monomers by enzymatic hydrolysis.
- In the second phase, acid-producing bacteria convert the simple structure of monomers into acetic acid, hydrogen and carbon dioxide. The acetic acid is a common by-product of digestion of fat, starch and proteins.
- The final phase is the methane-forming stage, where methanogenic bacteria play a role of converting a fermenting acetic acid into methane and carbon dioxide (Fig. 2).

Fig. 1 Cross section of
downdraft gasifier

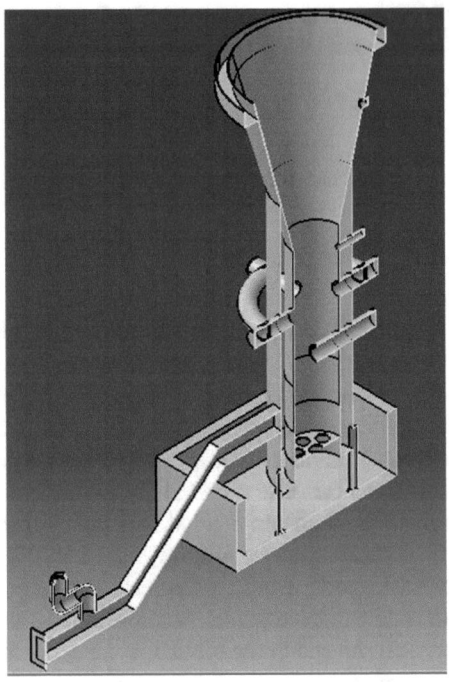

3 Poultry Litter as a Source

3.1 Proximate Analysis

The proximate analysis is used for finding how much percentage of moisture, volatile, fixed carbon and the ash content present in the sample. The sample considered here is the poultry litter. In this experiment, first we have ground the poultry waste into small particles for even heating. To find out the moisture content, the weighted sample is placed in the hot air oven, and the temperature is set to 110 °C. After the sample attains a certain temperature, the sample is allowed to cool for 15 min and again weighed. The reduction in the weight shows the moisture content of the poultry litter. Then the same sample is closed by a lid and moved into muffle furnace and heated at a temperature to 450 °C. The heated sample is allowed to cool for 10 min and is again weighed. The initial and final weights are compared to estimate the value of volatile content present in poultry litter. Then without any disturbance, place the same sample in muffle furnace with no lid and set a temperature of 750 °C. This step is carried out to find the ash content present in the sample. After the temperature reaches 750 °C, reduction in the mass of the sample is compared to the weight at 450 °C [11, 12]. This percentage shows the ash content present in the poultry litter. The proximate analysis results for poultry litter are given below (Fig. 3).

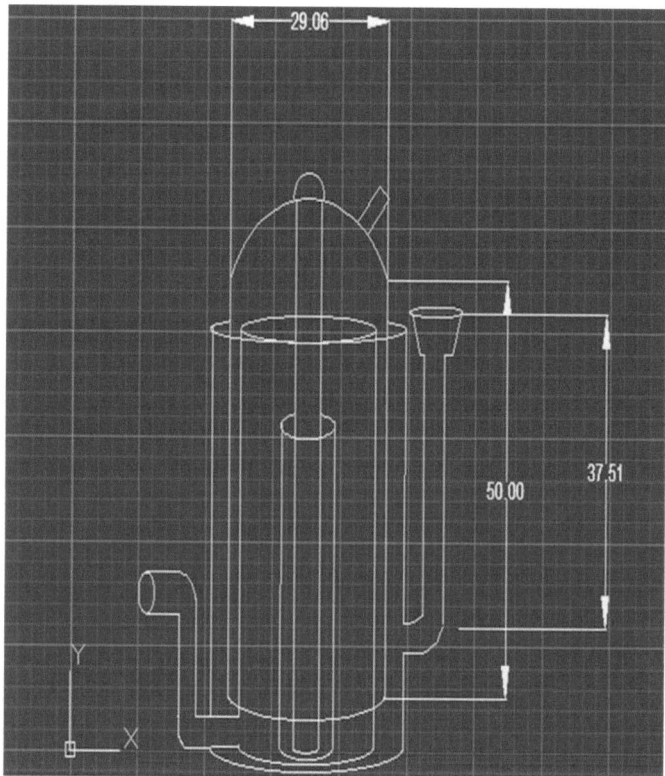

Fig. 2 AutoCAD design of nano-biogas set-up

% of moisture content = 14.80
% of volatile matter = 45.35
% of fixed carbon = 23.05
% of ash content = 16.80

Fig. 3 Placing the poultry sample in the muffle furnace for proximate analysis

3.2 Gasification

In this analysis, the downdraft gasifier is used which provides excellent handling for feeding biomass from top and the syngas extracts at the bottom of the gasifier with less tar content. The opening lid of the gasifier is insulated with water column for arresting the syngas, the bottom of the gasifier is insulated with water to prevent the escaping syngas, and ash is dropped on the water. For the combustion purpose, air is supplied to the gasifier by means of air blower with external power supply [13–15].

In the initial process, we brought the downdraft gasifier to combustion zone by burning some sawdust at 400 °C with sufficient air supplied through the blower. At this stage, the poultry litter is fed from top of the gasifier and air supply is reduced to 50%. This enhances the gasification and reduces the combustion. The lid of the gasifier is closed and filled with water column for arresting any gas leakages. Once the flue gas escapes from the outlet, then the syngas rushes out of the reduction zone. The resultant gas is highly combustible in nature. For analysing the present gas composition, the syngas is diverted to the producer gas analyser where the hose sucks up the syngas to the analyser and displays the readings in the LCD (Fig. 4).

In a downdraft gasifier, initially we add charcoal for red bed formation in the reduction zone. We use the fuel to ignite the charcoal, and the air is pumped using centrifugal pump for combustion process. Three thermocouples are inserted in the three zones—drying zone, distillation zone and reduction zone—for the purpose

2. <u>Downdraft or co-current gasifier</u>

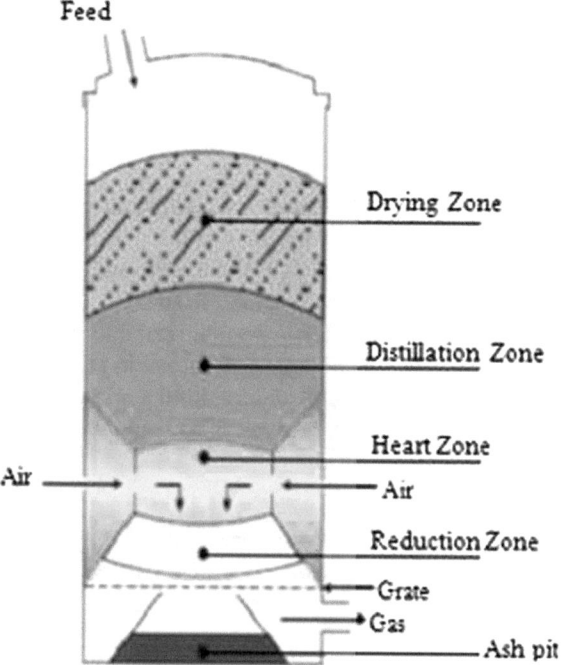

Feed

Drying Zone

Distillation Zone

Heart Zone

Air — Air

Reduction Zone

Grate

Gas

Ash pit

Downdraft gasifiers give relatively
cleaner gas (low tar content) and are
preferred for engine applications, niche
applications demanding clean gas.

Fig. 4 Operation of down draft gasifier

of monitoring the temperature. The poultry waste can be fed into gasifier when the temperature reaches 700 °C in the reduction zone. For the pyrolysis process, we allow the sub-stoichiometric air by the centrifugal pump. By allowing a partial amount of air into the gasifier, CO_2 formation in a reduction zone, CO and H_2 formation in pyrolysis zone and moisture content in drying zone can be reduced. The grate in the bottom of the gasifier should be adjusted periodically to remove the ash from the reduction zone, because ashes act as an insulator. By this process, the producer gas comes out of the gasifier and it is directed towards the producer gas analyser which it senses the producer gas and gives output about the composition about the CO, CO_2 and H_2. The peak values are given in Table 1.

Table 1 Peak value of producer gas

CO	48%
CO_2	28%
H_2	9%

3.3 Anaerobic Digestion of Poultry Litter

Anaerobic fermentation in the digester is a continuous process, which results in the formation of biogas. This experiment was conducted in KVIC biogas digester, which is suitable for biogas production in anaerobic environment. For effective digestion, the poultry litter should contain 10% solids. To attain this, the poultry waste is soaked in water and fed into the digester. After the retention time of about 15 days, the gas production starts from the digester. For measuring the composition of methane, carbon dioxide and hydrogen sulphide, the gas is sent to the biogas analyser, which sucks the biogas and shows the percentage of methane in the LCD. After the digestion of waste, the slurry is taken out of the digester (Table 2; Fig. 5).

Table 2 Peak value of biogas

CH_4	28%
CO_2	21%
H_2S	86 PPM

Fig. 5 Analysing the bio gas by using bio gas analyser

Fig. 6 After red bed
formation feeding poultry
waste into the gasifier

4 Work Profile

See Fig. 6.

5 Graphs and Results

Gasification of poultry litter produces more CO than the CH_4 produced through anaerobic digestion of the same. The CO_2 production in both the cases is relatively high. Therefore, the gasification process proves to be the most efficient method to produce energy from poultry litter waste (Figs. 7, 8 and 9).

6 Conclusion

Poultry waste directly disposed to agriculture land can increase nitrite content in soil which contaminates the groundwater. To address this issue and make revenue out of poultry waste, anaerobic digestion and gasification are suggested. In this paper, we made a comparative study of which method is efficient in terms of output. From

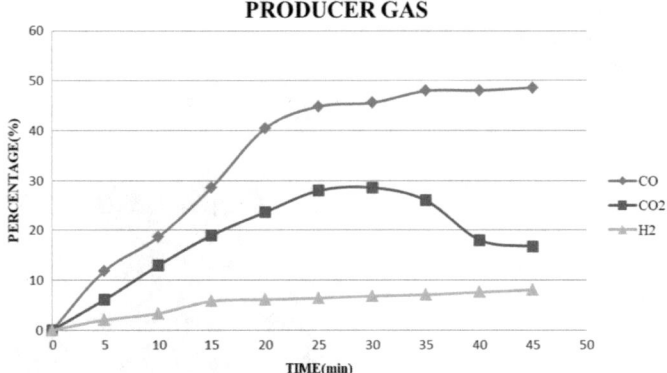

Fig. 7 CO, CO$_2$ and H$_2$ characteristic curve

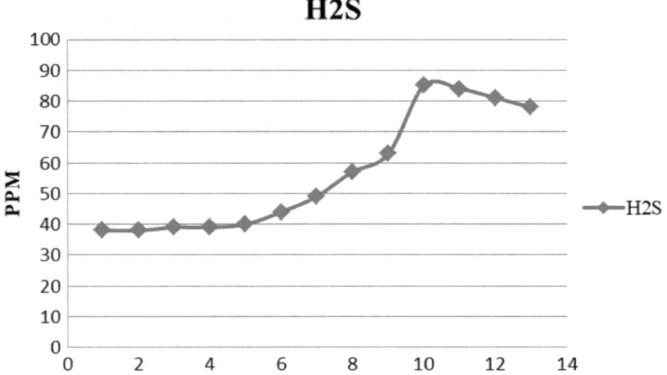

Fig. 8 H$_2$S characteristic curve with PPM

Fig. 9 CH$_4$ and CO$_2$
characteristic curve with
percentage

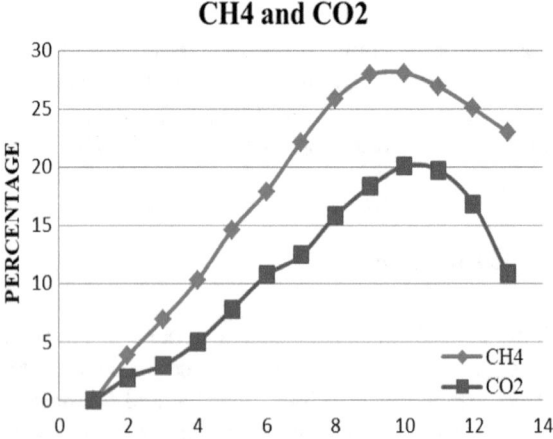

this analysis, the production of CO and H_2 is higher in gasification than the CH_4 produced through digestion. So we conclude that the gasification process is more efficient when compared to biogas process.

References

1. Khan, B. H. (2009). *Non conventional energy resources*. New Delhi: Tata McGraw-Hill.
2. Serio, M. A., Bassilakis, R., Kroo, E., & Wojtowicz, M. (2002). Pyrolysis processing of animal manure to produce fuel gases. *Fuel Chemistry Division Preprints, 47*(2), 588–592.
3. http://biomassproducer.com.au/producing-biomass/biomass-types/animal-waste/poultry-litter/#.W6jdU_kzbIU.
4. Integrated rural energy and waste management system through biogas technology, Grameen Shakti, Green Solutions, Dhaka, Bangladesh.
5. Markou, G. G. (2015). Improved anaerobic digestion performance and biogas production from poultry litter after lowering its nitrogen content. *Bioresource Technology, 196,* 726–730.
6. http://www.researchgate.net/publication/305570503_study_of_biogas_production_from_poultry_droppings_waste.
7. Miah, M. R. (2015). Concentrated their study on mixing the poultry waste with cow dung.
8. Saravanan, M. R., & Pasupathy, A. (2016). Incorporation of phase change material (PCM) in poultry Hatchery for thermal management & energy conversion schemes of slaughterhouse waste in Broiler farms for energy conservation—A case study. In *International Conference on Energy Efficient Technologies for Sustainability (ICEETS)*.
9. Petrecca, G., & Preto, R. (2011). The use of poultry manure to produce clean electric power. In *2011 International Conference on Clean Electrical Power (ICCEP)* (pp. 676–681).
10. Ushimaru, K. (2012). Sustainable green energy production from agricultural and poultry operations—A renewable energy project for community empowerment and vocational training in remote villages in South Africa. In *2012 IEEE Global Humanitarian Technology Conference* (pp. 310–314).
11. Kopeć, M., Gondek, K., Mierzwa-Hershtek, M., & Antonkiewicz, J. (2018, December). Factors influencing chemical quality of composted poultry waste Saudi. *Journal of Biological Sciences, 25*(8), 1678–1686.
12. Wang, L., Xue, B., & Yan, T. (2017, January). Greenhouse gas emissions from pig and poultry production sectors in China from 1960 to 2010. *Journal of Integrative Agriculture, 16*(1), 221–228.
13. Mau, V., & Gross, A. (2018, 1 March). Energy conversion and gas emissions from production and combustion of poultry-litter-derived hydrochar and biochar. *Applied Energy, 213,* 510–519.
14. Chaump, K., Preisser, M., Shanmugam, S. R., Prasad, R., & Higgins, B. T. Leaching and anaerobic digestion of poultry litter for biogas production and nutrient transformation. *Waste Management* (In press) (corrected proof, Available online 19 November 2018).
15. Heng, L. K. (2017, December). Bio gas plant green energy from poultry wastes in Singapore. *Energy Procedia, 143,* 436–441.

Study on Conversion Techniques of Alternative Fuels from Waste Plastics

Awinash Kumar, Santosh Kumar Dash, Moiching Sajit Ahamed and Pradip Lingfa

Abstract Plastic is one of the most useful, durable, cheap and ubiquitous materials known to human. It permeates every sphere of human life. Most of the developing countries are consuming more and more plastic goods due to modern urbanization and revenue upgrades. Plastic wastes are increasing day by day due to alternate production patterns and consumptions. Today plastic materials are produced twenty times more than fifty years ago. The most commonly used available methods for municipal waste in most of the countries is the dumping of waste plastic in open areas and landfills; which leads to an environmental threat. This study focused on the conversion techniques of alternative fuels from waste plastic materials. Conversion of plastic wastes can be regarded as a milestone of zero waste technology around the world. End use of plastic may be as a cheap source of energy. There are two approaches have been widely applied mechanical and chemical recycling. Altering of plastic wastes into energy resources approaches are pyrolysis with or without catalyst, gasification, and liquefaction. This paper also relates to another approach in conversion of plastic waste materials and by products like RDF (Refuse derived fuel) and RPF (Refused paper plastic fuel). These may be the next generation energy resources from waste plastic and municipal wastes.

Keywords Plastic wastes · Conversion techniques · Pyrolysis · Solid recovered fuels · International society of waste management · Air and water

1 Introduction

Need is the origin of an invention. Demand for the alternative sources of energy has been long realized. Though biofuel seems as a viable solution among many renewable energy sources to replace liquid petroleum diesel fuel, feedstock scarcity has become a major hindrance to its widespread use. Viscosity is not always reliable in case of biodiesel production. Liquid fuels flow behaviors have important role for

A. Kumar (✉) · S. K. Dash · M. S. Ahamed · P. Lingfa
Department of Mechanical Engineering, North Eastern Regional Institute of Science and Technology, Nirjuli, Itanagar, Arunachal Pradesh, India
e-mail: awinashwbut007@gmail.com

© Springer Nature Singapore Pte Ltd. 2020
S. K. Ghosh (ed.), *Energy Recovery Processes from Wastes*,
https://doi.org/10.1007/978-981-32-9228-4_18

their utilization [1–5]. In this regard; plastics derived liquid fuel explored by several researchers to reduce the pressure on diesel fuel and at the same time to control the issues pertinent to non-biodegradable plastic waste. Plastics were invented in the 1800s. Most of the developing countries are consuming more and more plastic goods due to modern urbanization and revenue upgrades. In between 17 years from 1990 to 2007 the production of plastic goods had reached 80 million tons to 260 million tons. According to a report of the government of India, import data of plastic goods mostly from countries like Canada, Denmark, Germany, U.K., Japan, Netherlands, France, and the USA during the years 1999 and 2000 is respectively more than 59,000 and 61,000 tons [6]. Plastic goods are becoming more popular day by day and plastic materials are widely used in house hold products, automobile industries, medical sectors, packaging of foods, electrical, mechanical industries as well as appliances and constructions of building or structures [7]. Plastic wastes are not easily flattened. Waste plastic disposal is a giant challenge for global environment. The world is now moving toward the zero-waste technology (ZWT) in accordance with green and safe ambience. Incineration and landfill are common and most popular way of decomposing wastes, but incineration of plastic wastes threat to the next pollution such as emission of CO_2 and soil pollution. As the plastics are made of polymers and they take several years in demolition due to their non-biodegradable nature. It's time to move forward waste to wealth technology (WWT). Energy from waste techniques plays a favorable role in the zero-waste technology a head.

With the expense of CH_4 and C_2H_4, more generation of CO and H_2 is obtained from the blends of polypropylene + polyethylene and polyethylene + polystyrene [8–10]. It is observed by the Shah SH, et al. that 275 °C is good enough for thermal degradation of PE for getting pyrolytic oil, wax, char, and pyrogas by 48.6, 10.1, 0.6, 40.1% respectively [11]. One of the conversion technique of polymer into liquid resources of fuel is liquefaction, occurs at the temperature range of 370–420 °C, the naphtha liquid is obtained by de-polymerization of polymers of waste plastics at 110 °C. Long chained hydrocarbons are in the range of C_6–C_{14} in naphtha liquid [12]. The experiment had been done by the Zen Feng et al. in 1996 indicates that in plastic liquefaction and coal-plastics co-liquefaction the nature of plastic and solvent have magnificent behavior on different catalysts added [13]. Many investigations have been done on the degradation of polyethylene and polypropylene are placed [14–25]. Most of the investigations made on PP, PE, and PS due to its favorable thermal degradation results for further analysis [26–30].

Gasification is also a method, defined as the thermo chemical conversion of the feedstock which is mainly carbon based material into combustible gas with the help of another gaseous compound. Gasification is useful for the heat treatment of pretreated heterogeneous waste and homogeneous carbon based wastes.

1.1 Refused Paper and Plastic Fuel (RPF): Application and Listing

RPF relates with refuse-derived paper and plastic-densified fuel. It is available in pellet form and bore range of 8–40 mm. It is a high calorie solid fuel made by conversion of the mixture of waste paper and plastics. In comparison with the coal it has the same energy level but is Eco-friendly and less CO_2 emission by 33%. The quantity and characteristics of MSW play an important role for the selection and utilization in different sectors [31].

1.1.1 Applications and Advantages of RPF

- The calorific value of RPF is adjustable for the requirement.
- It is economically efficient than using coal and in pellet form so easy way of transportation occurs.
- It can be a sustainable substitute for coal and coke at firms like paper industries, chemical manufacturers, and lime processing companies.
- It has high calorific fuel equivalent to coke.

To ensure consistent quality and to guarantee the reliability of the RPF, Japanese Industrial Standards Committee (JIS) have been established in 2006. Table 1 shows the standard value for the marketing aspects of the RPF.

1.2 Refuse-Derived Fuel (RDF): Fuel Resources Collected from Different Kinds of Non-biodegradable

RDF is a family of Solid recovered fuels (SRF). It is produced from municipal solid wastes (MSW), industrial wastes and it has high calorific value 18 MJ/kg. The methodology behind the production of RDF is separation techniques in which combustible matters are separated by non-combustible matters and then shredded.

Table 1 JIS standards for RPF (List number: JIS Z7311: 2010) [59]

RPF Type	RPF-coke	RPF			Unit
Grade	–	A	B	C	
Higher heating value	≥ 33	≥ 25	≥ 25	≥ 25	MJ/kg
	≥ 7883	≥ 5972	≥ 5972	≥ 5972	Kcal/kg
Moisture	$3\leq$	$5\leq$	$5\leq$	$5\leq$	%
Ash	$5\leq$	$10\leq$	$10\leq$	$10\leq$	%
Residual chlorine	$0.6\leq$	$0.3\leq$	$0.3 \leq 0.6 \leq$	$0.6 \leq 2.0 \leq$	%

Magnetic separation and eddy current separation is generally used. It is available in pellet, bricks or logs forms. The vital problem is explosion during production (shredding) of RDF and it may be prevented by explosion-suppression systems. To ensure the quality production of SRF one such standard is European standard CEN/34 Kilns. At present the UK is a big exporter of RDF about 2.5MT/year to Europe [32]. During the combustion of RDF the NO_X and SO_2 emission concur with the limits as per the European Union directives 2000/76/EC, was found. International Energy Agency and International Solid Waste Association member countries are looking through the possibilities for trending waste energy product as RDF. The countries like Italy, Denmark, Belgium, and Netherland are using at least one cement kiln processing RDF. In EU the application of RDF in various cements factories since 1993 [31, 33–36].

1.3 Target and Sources of Plastic Wastes

The market research is essential to storing up large quantities of plastic to ensure that the recycling process of will run smoothly. Potential of plastic waste should coordinate with the marketing revenue. Rejected or waste plastic materials from manufacturing industries of plastics goods have better characteristics for recycling or producing fuel resources. Agricultural wastes such as plastic bags, plastic containers, hosepipes, sheets may be reused as fuel resources in the way of zero-waste technology (ZWT). A lot of plastics generally polyethylene (PE) can be obtained from craftsmen shops and supermarkets. On the other hand municipal waste is also a vital source of plastic wastes. The local scrap yards of cities have sufficient broken plastic furniture those have been collected by scrap buyer moving around the locality. Water bottles can be collected from nearby the major connecting railway stations especially at junctions. Common sources of plastic wastes are pharmaceutical or clinical, industrial, construction and electronic wastes (EW) [37].

1.4 Pre-use Plastic (Production Scrap)

Pre-use plastic waste is likely either to be plastic. These have not met the specification required for their designed use, or off-cuts arising during assembly or installation. Although these materials are not suitable for intend uses, it may be suitable for other applications and have the potential to be recycled. This type of plastic waste is the main source of plastics suitable for reprocessing from manufactures of plastic products. It is typically more valuable than post-use plastic wastes. It generally requires little processing to use in a new product [38–40].

1.5 Post-Use Plastic

Recycling categories of Post-use plastics:

- Water Bottles, pots, tubes and trays
- Plastic films
- Pipes and Rigid plastics
- An expanded polystyrene (EPS)
- Flexible plastics, such as strapping and cable sheathing (LDPE).

1.6 Selection of Waste Plastics: A Promising Resource of Fuel Production

It is investigated by many researchers that the conversion of waste plastic into fuel requires combustible and nonhazardous feed stock. The conversion methodology of plastic to fuel depends upon the types of plastic chosen. A separation process is also done for the identification of different types of plastics. The tests by water, burning, smell after burning, scratch is done for the proper identification. All plastics are not suitable for recycling. Thermo sets or thermosetting can be heated and take shape only once [41, 42].

These types of polymers are not suitable for repeated thermal processes. Thermoplastics (PE, PVC, PP,) are good source of fuel. Polytetrafluoroethylene (PTFE) is also a type of thermoplastic. Almost following types of plastics which are recycled:

- High-density polyethylene (HDPE)
- Low-density polyethylene (LDPE)
- Polystyrene (PS)
- Polypropylene (PP)
- Polyvinyl chloride (PVC).

1.7 Plastic to Fuel Techniques: Recycling of Waste Polymers

Recycling of plastic wastes is a necessary step toward the promotion of a sustainable environment. In general it was found following four ways of recycling of plastics:

- Primary Recycling—altering of plastic wastes into products having performance level comparable to that of original products made from virgin plastics. It is expensive. It is a form of mechanical reprocessing in which product has the similar specifications as that of virgin. Semi clean scraps may have good results with this process. It is not an appropriate process.

- Secondary Recycling—altering of plastic wastes having less demanding performance requirements than the original material. This process is done for wealth from waste. The process involves different products in comparison with the original one.
- Tertiary Recycling—It is the most common process used in the field of plastic to fuel technology. Chemicals/fuels/ similar products from plastic wastes are produced by de-polymerization or thermal degradation. The raw material loses its original value. In this process, high level of waste contaminants can be achieved by thermal cracking.
- Quaternary Recycling—Incineration play important role for energy recovery process from plastic wastes. Incineration is caused by the generation of energy in this process. This process is more convenient for solving air pollutions not suitable for solid waste problems. As 20% by weight and 10% by volume of the virgin wastes are land filled [43, 44]. Figure 1 shows the general process for making liquid fuel from waste plastic materials.

1.8   Pyrolysis (with or Without Catalyst): Sustainable Zero-Waste Technology

Pyrolysis is an endothermic process can be done with or without catalyst. Table 2 shows the conversion rates of different types of resins. Conversion parameters are necessary for selection of resin before pyrolysis. The operating temperature range is 350–600 °C. Also an investigation report of State Research, Inc. (NSR) founded that the suitable and feasible way for conversion of polymer into fuel resources is pyrolysis

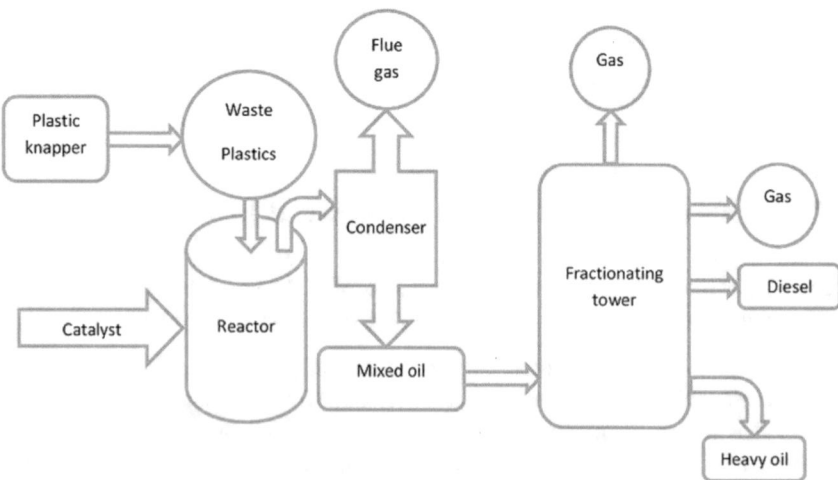

Fig. 1 Generation of fuel from the plastic wastes [58]

Table 2 Conversion rates of some resins and their sources [60]

Resin	Availability and identification	Conversion rate (%)
HDPE	Milk jugs, cleaning agents, laundry detergent, bleaching agents, shampoo bottles	Not available
LDPE	Crushed bottles, shopping bags, most of the wrappings and high resistance sacks	70
PS	Toys, hard packing, refrigerator trays, cosmetic bags, costume jewelry, audio Cassettes, CD cassettes	80–85
PP	Polypropylene, furniture consumers, luggage, toys, bumpers, lining and external borders of cars	50–60
PET	Soft drinks bottles, mineral water bottles, juice containers, cocking Oil packs	30
PVC	A tray for sweets, fruits, plastic packing, and foils to wrap the foodstuff	30

without any catalyst [45]. The product output by simple pyrolysis as a fluid fuel has impurities like chlorine, sulfur and residues, which may be optimized by using a catalyst. Pyrolysis forms three products like charcoal, condensable liquids (tar) and non-condensable gases. By using catalyst, pyrolysis is done with these types of plastic wastes, suitable for conversion of wastes of plastic into energy resources. There are four types of kinetics under pyrolysis known as mechanism of thermal degradation. Proposed mechanisms are: (1) end-chain scission, (2) fragmentation, (3) elimination of side chain, (4) cross-linking. For thermosetting plastics cross linking mechanism is preferred. It depends on the types of polymer [46, 47]. Fujimoto [47] and Panda et al. [6] suggested Fe/activated carbon catalyst for the catalytic degradation of PP through the initiation of random breakage of C-C bond followed by propagation and termination. Polyethylene (PE) type plastics like HDPE, LDPE and PP have crossed chain hydrocarbon chemical structure, so conducting pyrolysis is somehow difficult [48]. Pyrolysis with catalysts such as Red Mud, FCC, HZSM-5, Y-zeolite, Fe_2O_3, Al_2O_3, $Ca(OH)_2$ and natural zeolite have been used for further improvement of liquid by product obtained from it. In thermal degradation the role of catalyst is caused for better fraction. Also less energy consumption was founded in thermo-catalytic than the thermal degradation [49]. MCM-41 and ZSM-5 zeolite catalyst in the staged pyrolysis of plastic wastes is more convenient for the gasoline. Table 3 shows which types of the resin is as suitable as for the pyrolysis. Liquid hydrocarbon products may be favorable for fuel production due to low sulfur content. Thermal cracking involves the chain scission of polymers to produce basic petrochemical compounds at different temperatures, reaction times, pressures, in the presence or absence of reactive gases and catalysts. Supply of heat is vital for breaking of the bonds and results in the initiation of a volatile fragment and a carbonized char, further separated into non-condensable gases and condensable hydrocarbon oil. Recovery of various hydrocarbon fragments in the fuel range has been focused by the chemical recycling, thermolysis, thermo catalytic cracking and steam decomposition of solid plastic wastes [45, 50–52]. Thermal pyrolysis conducts to a wide range of C_5–C_{80}

Table 3 Outcomes of some plastic pyrolysis [49]

Products	Plastics	Suitable feed stock for liquid fuel
Liquid hydrocarbons	High-density polyethylene (HDPE)	Yes
	Low-density polyethylene (LDPE)	Yes
	Polystyrene (PS)	Yes
	Polypropylene (PP)	Yes
	Polymethylmethacrylate (PMMA)	Yes
	Acrylonitrile-butadiene-styrene copolymer (ABS)	No
Solid products	Polypropylene terephthalate (PET)	No
Carbonous products	Polyurethane (PUR)	No
	Phenol resin (PF)	No
Hydrogen chloride and carbonous products	Polyvinyl chloride (PVC)	No
	Polyvinylidene chloride (PVDC)	No

hydrocarbons. The limited commercial value products are produced like I-olefin and n-paraffin at high temperature. Heavy oils are obtained in form of wax. The key drawback of this process is its wide range of production of hydrocarbons at about 500–900 °C. Thermal pyrolysis may be done in two step process also [53–55]. Thermal cracking is an emerging technique for converting waste plastics into source of energy [56, 57].

Following advantages of pyrolysis process: [46]

- Low capital cost is obtained.
- First and easy process to recovering energy resources from waste.
- It supports zero-waste technology.
- It reduces volume of waste in the range of 50–90%.
- Easy to transport and store.
- A desirable method to produce revenue from wastes.

For lighter hydrocarbons high reaction temperature is needed, it is the major drawback of thermal degradation [7].

1.9 Gasification: A Thermo Chemical Conversion

Gasification may be called as a first order reaction occurs in between temperature range greater than 600–800 °C usually in less oxygen environment. Gasification may be endothermic or exothermic in practice. Syngas or producer gas is an output by product of gasification [8]. Under the gasification process 1.76 MJ/m^3 of energy and cold gas efficiency approx. 66% was obtained from 55% of plastic wastes with a refuse-derived fuel briquette. Gasification is about to partial oxidation of organic matter for the production of syngas. At the first stage, partial combustion of biomass takes place to produce gas and char. Heterogeneous reactions take place in the presence of gasification agent [49].

There are three main components of gasification systems:

- The gasifier; it helps to produce combustible gas.
- The gas cleaning arrangements; it removes harmful compounds or undesirable compounds from the combustible gas.
- The energy recovery arrangement; it recovers the desired energy from the process.

2 Conclusions

The study reveals that research and investigation efforts on the conversion techniques using different catalysts and agents have been placed. So many research works have been initiated but there is limited research about to marketing aspects of plastic fuel. Laboratory to land terminology is most important for any research work. Awareness campaigns will lead to growth of plastic oil industry.

There are following conclusions:

(1) It is obvious that most of the research works regarding conversion of plastic wastes into hydrocarbon products are based on thermal cracking or pyrolysis (with or without catalyst). Not a single appropriate approach for conversion is identified till now.
(2) Limited research works are available on SRF, RPF and RDF.
(3) Limited works have done on the IC engine performance with plastic oil blends and physiochemical properties of plastic oil.
(4) Government policies in India and standardization of plastic fuel are not fixed like bio fuels.
(5) The issues on scaling and sustainability of industries like plastic to fuel in world wide.
(6) Cost competitiveness with normal fuel and marketing research will reach to acceptance of plastic fuel.

References

1. Dash, S. K., & Lingfa, P. (2017). A review on production of biodiesel using catalyzed transesterification. In *AIP Conference Proceedings* (p. 1859. 020100).
2. Dash, S. K., & Lingfa, P. (2018). An overview of biodiesel production and its utilization in diesel engines. In: *IOP Conference Series: Material Science and Engineering* (Vol. 377, p. 012006).
3. Dash, S. K., & Lingfa, P. (2018). Performance evaluation of Nahar oil-diesel blends in a single cylinder direct injection diesel engine. *International Journal of Green Energy, 15,* 400–405.
4. Dash, S. K., & Lingfa, P. (2018). Production of biodiesel from high FFA non-edible Nahar oil and optimization of yield. In *Advanced Manufacturing of Material Sciences* (pp. 431–439).
5. Kumar, A. (2015, May–June). Debottlenecking of Bernoulli's apparatus and verification OF Bernoulli's principle. *IOSR Journal of Mechanical and Civil Engineering, 12*(3), 116–146. e-ISSN: 2278-1684, p-ISSN: 2320-334X (Ver. II).
6. Panda, A. K., Singh, R. K., & Mishra, D. K. (2010). Thermolysis of waste plastics to liquid fuel. A suitable method for plastic waste management and manufacture of value added products—A world prospective. *Renewable and Sustainable Energy Reviews, 14*(1), 233–248.
7. Mohanraj, C., Senthilkumar, T., & Chandrasekhar, M. (2017, January 2) A review on conversion techniques of liquid fuel from waste plastic materials. *International Journal of Energy Research.*
8. Brems, A., Dewil, R., et al. (2013). Gasification of plastic waste as waste-to-energy or waste-to-syngas recovery route. *Natural Science*, 695–704.
9. Lopez, A., De Marco, I., Caballero, B. M., Laresgoiti, M. F., Adrados, A., & Torres, A. (2011). Pyrolysis of municipal plastic wastes II: Influence of raw material composition under catalytic conditions. *Journal of waste management, 3,* 1973–1983.
10. Wilk, V., & Hofbauer, H. (2013). Conversion of mixed plastic wastes in a dual fluidized bed steam gasifier. *Journal of Fuel*, 787–799.
11. Shah, S. H., Khan, Z. M., Raja, I. A., Mahmood, Q., Bhatti, Z. A., Khan, J., et al. (2010). Low temperature conversion of plastic waste into light hydrocarbons. *Journal of Hazardous Materials, 179,* 15–20.
12. Biofuel Academy, Liquefaction of plastic waste.
13. Feng, Z., et al. (1996). Direct liquefaction of waste plastics and coliquefaction of coal-plastic mixtures.
14. Kim, J. R., Yoon, J. H., & Park, D. W. (2002). Catalytic recycling of the mixture of polypropylene and polystyrene. *Polymer Degradation and Stability, 76,* 6167.
15. Luo, K., Suto, T., Yasu, S., & Kato, K. (2000). Catalytic degradation of high density polyethylene and polypropylene into liquid fuel in a powder-particle fluidized bed. *Polymer Degradation and Stability, 70,* 97–102.
16. Serrano, D. P., Aguado, J., Escola, J. M., & Garagorri, E. (2001). Conversion of low density polyethylene into petrochemical feedstocks using a continuous screw kiln reactor. *Journal of Analytical and Applied Pyrolysis, 58–59,* 789–801.
17. Hwang, E. Y., Kim, Y. R., Choi, J. K., Woo, H. Y., & Park, D. W. (2002). Performance of acid treated natural zeolites in catalytic degradation of polypropylene. *Journal of Analytical and Applied Pyrolysis, 62,* 351–364.
18. Walendziewski, J., & Steininger, M. (2001). Thermal and catalytic conversion of waste polyolefins. *Catalysis Today, 65,* 323–330.
19. Uddin, A., Koizumi, K., Murata, K., & Sakata, Y. (1997). Thermal and catalytic degradation of structurally different types of polyethylene into fuel oil. *Polymer Degradation and Stability, 56,* 37–44.
20. Breen, C., Last, P. M., Taylor, S., & Komadel, P. (2000). Synergic chemical analysis the coupling of TG with FTIR, MS and GC-MS 2. Catalytic transformation of the gases evolved during the thermal decomposition of HDPE using acid-activated clays. *Thermochimica Acta, 363,* 93–104.
21. Miranda, R., Yang, J., Roy, C., & Vasile, C. (2001). Vacuum pyrolysis of commingled plastics containing PVC I. Kinetic study. *Polymer Degradation and Stability, 72,* 469–491.

22. Ukei, H., Hirose, T., Horikawa, T., Takai, Y., Taka, M., Azuma, N., et al. (2000). Catalytic degradation of polystyrene into styrene and a design of recyclable polystyrene, with dispersed catalysts. *Catalysis Today, 62,* 67–75.
23. Murata, K., Hirano, Y., Sakata, Y., & Uddin, A. (2002). Basic study on a continuous flow reactor for thermal degradation of polymers. *Journal of Analytical and Applied Pyrolysis, 65,* 71–90.
24. Ali, S., Garforth, A. A., Harris, D. H., Rawlence, D. Y., & Uemichi, Y. (2002). Polymer waste recycling over "used" catalysts. *Catalysis Today, 75,* 247–2–255.
25. Grieken, R., Serrano, D. P., Aguado, J., Garcia, R., & Rojo, C. (2001). Thermal and catalytic cracking of polyethylene under mild conditions. *Journal of Analytical and Applied Pyrolysis, 58–59,* 127–142.
26. Karaduman, A., Sximsxek, E. H., Cicek, B., & Bilgesu, A. Y. (2001). Flash pyrolysis of polystyrene wastes in a free-fall reactor under vacuum. *Journal of Analytical and Applied Pyrolysis, 60,* 179–186.
27. Karaduman, A., Sximsxek, E. H., Cicek, B., & Bilgesu, A. Y. (2002). Thermal degradation of polystyrene wastes in various solvents. *Journal of Analytical and Applied Pyrolysis, 62,* 273–280.
28. Gao, Z., Amasaki, I., Kaneko, T., & Nakada, M. (2003). Calculation of activation energy from fraction of bonds broken for thermal degradation of polyethylene. *Polymer Degradation and Stability, 81,* 125–130.
29. Faravelli, T., Bozzano, G., Colombo, M., Ranzi, E., & Dente, M. (2003). Kinetic modeling of the thermal degradation of polyethylene and polystyrene mixtures. *Journal of Analytical and Applied Pyrolysis, 70,* 761–777.
30. Faravelli, T., Pinciroli, M., Pisano, F., Bozzano, G., Dente, M., & Ranzi, E. (2001). Thermal degradation of polystyrene. *Journal of Analytical and Applied Pyrolysis, 60,* 103–121.
31. Pattnaik, S., & Reddy, M. V. (2009). Assessment of municipal solid waste management in Puducherry (Pondicherry), India. *Resources, Conservation and Recycling, 54*(8), 512–520.
32. Vijaykumar, B., Chandanashetty, & Patil, B. M. (2015). *International Journal on Emerging Technologies* (special Issue on NCRIET-2015), 121–128.
33. Junior, L. M. (2003). *Sustainable development and the cement and concrete industries* (Ph.D. thesis). Universite de Sherbrooke, Quebec.
34. Rogaume, T., Auzanneau, M., Jabouille, F., Goudeau, J. C., & Torero, J. L. (2002). The effects of different airflows on the formation of pollutants during waste incineration. *Fuel, 81,* 2277–2288.
35. Gendebien, A., Leavens, A., Blackmore, K., Godley, A., Lewin, K., Whiting, K. J., et al. (2003). Refuse derived fuel, current practice and perspectives. Final Report, European Commission.
36. Liu, W., Hu, C., Yang, Y., Tong, D., Li, G., & Zhu, L. (2010). Influence of ZSM-5 zeolite on the pyrolytic intermediates from the co-pyrolysis of pubescens and LDPE. *Energy Conversion and Management, 51*(5), 1025–1032.
37. ACRR. (2004). *Good practices guide on waste plastics recycling.* Brussels, Belgium: Association of Cities and Regions for Recycling.
38. Aguado, J., Serrano, D. P., & Miguel, G. (2007). European trends in the feedstock recycling of plastic wastes. *Global NEST Journal, 9,* 12–19.
39. Garforth, A., Ali, S., Hernandez-Martinez, J., & Akah, A. (2004). Feedstock recycling of polymer wastes. *Current Opinion in Solid State and Materials Science, 8,* 419–425.
40. Kyrikou, I., & Briassoulis, D. (2007). Biodegradation of agricultural plastic films: A critical review. *Journal of Polymers and the Environment, 15,* 125–150.
41. Morris, J. (1996). Recycling versus incineration: An energy conservation analysis. *Journal of Hazardous Materials, 47,* 277–293.
42. Swift, G., & Wiles, D. (2004). Degradable polymers and plastics in landfill sites. *Encyclopedia of Polymer Science and Technology, 9,* 40–51.
43. Bandopadhyay, T. K., & Sharma, S. (2004, August–October). Management of plastics, polymer wastes and bio-polymers and impact of plastics on the eco-system. 2(5).
44. Patni, N., Shah, P., Agarwal, S., & Singhal, P. (2013, April 28). Alternate strategies for conversion of waste plastic to fuels. *ISRN Renewable Energy, 2013.*

45. Silvério, F. O., Barbosa, L. C. A., & Piló-Veloso, D. (2008). A pirólise como técnica analítica. *Quimica Nova, 31*(6), 1543–1552.
46. Singhad, R. P., Tyagib, V. V., Allen, T., et al. (2011). An overview for exploring the possibilities of energy generation for municipal solid waste (MSW) in Indian scenario. *Renewable and Sustainable Energy Reviews, 15*(9), 4797–4808.
47. Sekine, Y., & Fujimoto, K. (2003). Catalytic degradation of PP with an Fe/activated carbon catalyst. *Journal of Material Cycles and Waste Management, 5*(2), 107–112.
48. Di Blasi, C. (2000). Dynamic behaviour of stratified downdraft gasifier. *Chemical Engineering Science, 55*(15), 2931–2944.
49. Beyene, H. D. (2014). Recycling of plastic waste into fuels, a review. *International Journal of Science and Society, 2*(6), 190–195.
50. Al-Salem, S. M., Lettieri, P., & Baeyens, J. (2009). Recycling and recovery routes of plastic solid waste (PSW): A review. *Waste Management (New York, N.Y.), 29*(10), 2625–2643.
51. Lee, K.-H. (2006). Thermal and catalytic degradation of waste HDPE. In J. Scheirs & W. Kaminsky (Eds.), *Feedstock recycling and pyrolysis of waste plastics* (pp. 129–160). Hoboken: Wiley.
52. Zhang, G.-H., & Zhu, J.-F. (2006). Prospect and current status of recycling waste plastics and technology for converting them into oil in China, pp. 231–239.
53. Aguado, J., Serrano, D. P., & Escola, J. M. (2006). Catalytic upgrading of plastic wastes. In J. Scheirs & W. Kaminsky (Eds.), *Feedstock recycling and pyrolysis of waste plastics* (pp. 73–110). Hoboken: Wiley.
54. Marcilla, A., Beltrán, M. I., & Navarro, R. (2009). Thermal and catalytic pyrolysis of polyethylene over HZSM5 and HUSY zeolites in a batch reactor under dynamic conditions. *Applied Catalysis, B: Environmental, 86*(1–2), 78–86.
55. Htet, M. T. (2010, June). *Recycling of waste plastic in a two step thermo-catalytic reaction systems* (Ph.D. thesis).
56. Demirbas, A. (2004). Pyrolysis of municipal of plastic wastes for recovery of gasoline-range hydrocarbons. *Journal of Analytical and Applied Pyrolysis, 72*(1), 97–102.
57. Abbas-Abadi, M. S., Haghighi, M. N., & Yeganeh, H. (2012). The effect of temperature, catalyst, different carrier gases and stirrer on the produced transportation hydrocarbons of LLDPE Degradation in a stirred reactor. *Journal of Analytic and Applied Pyrolysis, 95,* 198–204.
58. Website: http://www.oceanrecov.org. Valuing Plastic.
59. Website: http://www.ecomining.co.jp.
60. Zhang, G. H., Zhu, J. F., & Okuwaki, A. (2007). Prospect and current status of recycling waste plastics and technology for converting them into oil in China. *Conservation & Recycling, 50*(3), 231–239.

Co-processing of RDF in Cement Plants

Kaushik Chandrasekhar and Suneel Pandey

Abstract Effective solid waste management in India is becoming a major need due to high pace of urbanization and rapid industrialization posing major challenges. Co-processing of refuse-derived fuel (RDF) presents to the urban local bodies an option for environmentally sound management for non-biodegradable, non-recyclable organic waste as well as opportunity to reduce emissions due to fossil fuels used in cement kilns. With about 68.8 Million Tonnes of MSW generated annually in the country and 13.7 Million Tonnes available for co-processing, the potential for co-processing looks positive. The government has also realized the potential and introduced policies to encourage the development of co-processing. However ground realization of this potential still remains in the nascent stage with Thermal Substitution Rate (TSR) reported at less than 1%. Municipal Solid waste is still considered a burden and a cost for the municipalities which spend large portions of its budget to transport waste to landfills. The quality of RDF produced has had an impact on the levels of utilization in the cement kilns and the associated environment impact. Currently, RDF due to the lack of quality consideration, is finding little or no its use in the cement plants. The informal sector could play a key role in assisting to manage collection of wet and dry waste and manning Material Recycling Facilities. However, the integration of this sector formally into the solid waste management system has been a hiccup. The paper presents analysis of factors leading to low levels of co-processing in cement plants despite its advantages, national and regional policies, economic feasibility, industry perspective and proposes menu of business models for turning co-processing into a sustainable business case.

Keywords Solid waste management · Co-processing · Refuse-derived fuel · Business models

K. Chandrasekhar (✉)
Centre for Waste Management, TERI, Bangalore, India
e-mail: k.chandrasekhar@teri.res.in

S. Pandey
Centre for Waste Management, TERI, New Delhi, Delhi, India
e-mail: spandey@teri.res.in

© Springer Nature Singapore Pte Ltd. 2020
S. K. Ghosh (ed.), *Energy Recovery Processes from Wastes*,
https://doi.org/10.1007/978-981-32-9228-4_19

1 Introduction

Accelerated urbanization has been one of facets of developing India and has led to inadequate services, infrastructure and sanitation in the urban areas. With 410 million urban dwellers India along with China account to 30% of world's urban population. Between 2014 and 2050, India is expected to add another 404 million people to its urban community.[1] This growth and increase in population density often creates stress on the basic civic amenities offered by local governments. The fourteenth finance commission report on municipal finances and service delivery in India observed that due to the widening gap in the providing basic services one of the priority areas for investment of grant recommended by 14FC was services such as water, sewerage, storm water and sanitation.[2] Today, India generates close to 144,165 TPD[3] of waste with an organic fraction of 40–60%,[4] recyclable fraction of 10–25% and moisture content of 30–60%. CPCB estimates that 28.4% is treated with collection efficiency of 80.28% and the remaining is disposed in landfills or open dumps. The low treatment efficiency has created an increased load on landfill sites eventually converting them to dump sites. While weak business models and lack of segregation has hampered traditional waste processing technologies such as composting, ULB's are always looking at low cost, effective strategies to treat and dispose waste generated.

The growth in GDP and urban agglomerations has also led to increase in demand for the construction industry. Cement being the heart of the construction industry is an energy intensive industry with power and fuel cost accounting to 30–35% of the overall production cost.[5] Cement production is also responsible for 5% of the global anthropogenic CO_2 emissions and 7% of industrial fuel use.[6] Thus the industry is an important contributor to GHG emissions and reductions measure in this sector will lead to significant reduction in overall GHG emissions.

Coal is the major source for energy fuelling the cement sector alongside petcoke. The Indian cement industry continues to suffer on account of reduced availability of linked coal. CMA reported coal receipt against FSA/Linkage by member units as 7.06 Mn tonne in 2015–16 as against 7.71 Mn in 2014–15 and 9.22 Mn Tonne in 2013–14.[7] This coal receipt in 2015–16 formed only 24.3% of the total fuel consumption by the industry. Due to this shortage and increased demand for cement, the industry is required to buy coal from e-auctions, import coal or look for alternate fuel sources like husk, MSW and biomass. The co-processing of MSW based RDF as an alternate

---

[1] UN World Urbanization Prospects—2014.

[2] Municipal Finances and Service delivery in India, Government of India.

[3] Central Pollution Control Board [1].

[4] Assessment of Status of Municipal Solid Wastes Management in Metro Cities and State Capitals. Retrieved from http://www.cpcb.nic.in/wast/municipalwast/Studies_of_CPCB.pdf on 29.08.2016.

[5] Crisil Industry Research Sample Report https://www.crisil.com/Crisil/pdf/research/industry-research-sample-report.pdf.

[6] Boesch and Hellweg [2].

[7] Cement Manufacturers Association Annual report 2014–15 and 15–16.

fuel in cement kilns has shown positive potential in replacing conventional fuel and addressing the solid waste management concerns. Co-processing is a scientifically proven and established technology for disposal of hazardous and other non-recyclable waste in an environmentally sustainable manner.[8] As the process involves waste combustion with no residual waste left over, co-processing ensures complete waste disposal as compared to incineration. Further, every ton of RDF could yield 1.48–1.88 ton CO_2 emission reductions[9] due to avoided coal.

The current paper starts by exploring RDF as a potential fuel and its associated benefits. This is followed by a review of reasons leading to lower levels of TSR. This is further validated by an analysis of effective availability of waste for RDF co-processing and a meta-analysis of literature on quality standards of RDF for co-processing and integration of informal sector. Based on these assessments the paper further discusses the need for sound business models for ensuring sustainability of the process and suggests a few technical models for implementation in the Indian context.

2 Methodology for Review

The study involves a literature review of acceptable quality standards of RDF for co-processing and need for integration of informal sector. The following was the assessment methodology applied for analysis of effective availability of waste.

An area of 100 km radius around Cement plants in major cement producing states of India contributing to over 75% of production of cement[10] was mapped. Cities having a municipal solid waste generation of more than 100 TPD were mapped to track waste contributions. Waste availability for producing RDF was assumed at 20% factoring informal sector activity, collection efficiency and competition from waste to energy plants for RDF. The yield from RDF manufactured from MSW was assumed to be 10%. The calorific value of resultant RDF was assumed to be 2500 kcal/kg.

3 Potential for RDF Use as a Fuel

A study done for evaluating the potential environment impacts due to use of alternate fuel such as RDF, Tire Derived Fuel and biological sludge or a mixture of them in cement plants by partial replacement of conventional fuel observed that RDF provided the most environment friendly prospect as compared to the other three fuels.[11] Replacing coal and petcoke with lower carbon alternatives has always been on

---

[8]Promoting alternate fuel and raw material usage in Cement Industry—CII and Shakti Foundation.

[9]Industrial Efficiency Technology Database—RDF Co-processing.

[10]Cement Manufacturers Association—2011–12 data.

[11]Georgiopoulou and Lyberatos [3].

Table 1 CO_2 emissions from fuel sources

Fuel type	Petcoke	Coal	Natural gas	Tires	Waste oil	Plastic	MSW
Net Co_2 emission factor (Kg Co_2/GJ)	101	96	54.2	85	74	75	8.7

the agenda for cement industries. The International Energy Agency (IEA) observes that most of the alternate fuels are 20–25% less carbon intensive than traditional fossil fuels.[12] Table 1[13] shows the potential for AFR to reduce GHG emissions in the cement production process.

At the current rate of waste generation, 1,557,558[14] tons of carbon reductions can be realized if solid waste generated in India could be converted into RDF and substituted for coal, contributing to the Indian Nationally Determined Contributions (NDC's). As per CII, thermal substitution rate (TSR) by AFR use in cement industry is anticipated to reach 25% by 2025 and RDF use is going to contribute 57.06% of the substituted AFR[7] provided use of RDF as alternate fuel is promoted in cement kilns. The Government of India (GoI) policies have endorsed the use of RDF in cement plants in a positive way.[15] The government has introduced the definition of co-processing under Solid Waste Management rules (SWM), 2016. Through the rules it has further stated that non-recyclable, high calorific value fractions of waste are to be segregated and sent to waste to energy, RDF production, co-processing in cement plants or thermal power plants. The rules further reinforce that industrial units located within one hundred km from RDF plants or waste to energy plants based on solid waste should make arrangement within six months of notification to replace at least five per cent of fuel requirement by RDF. Further the GoI has come up with the draft guidelines for pre-processing and co-processing of hazardous and other wastes in cement plants.[16]

Considering the aforementioned potential and positive impacts that RDF co-processing could generate the proportional replication or thermal substitution rate in cement plants has been considerably low at 3–4%[17] while in few countries it has been as high as 60%. Some reasons for these lower levels of TSR substitution include.

---

[12]Cement Technology Roadmap 2009.

[13]Adapted from Alternative Fuel Use in Cement Manufacturing, The Pembina Institute and Environmental Defence.

[14]Source: TERI research; Assuming daily waste generation as per CPCB 2014 estimates and 20% recovery efficiency and 10% yield; 1.48 ton CO_2 emission reduction per tonne usage of RDF.

[15]Solid waste Management Rules, 2016, Ministry.

[16]Draft Guidelines for Pre-processing and Co-processing of Hazardous and Other Wastes in Cement Plants as per H&OW(M&TBM) Rules 2016—http://www.cpcb.nic.in/final_report_27.01.17.pdf.

[17]CMA Annual report 2015–16.

4 Availability of Acceptable Quality RDF for Co-processing

RDF typically is manufactured in the form of fluff, pellets or bricks and consists of segregated municipal waste including organic waste and materials like paper, cloth, plastic and wood that provides the fuel calorific value to burn.[18] However, as global waste hierarchy defines recycling with a higher precedence over utilization of waste for producing energy. Therefore non-recyclable, non-biodegradable organic waste would constitute RDF.[19] This would be mean soiled organic waste too dirty to recycle, rejects from composting and anaerobic digestion could be used to process RDF. However, RDF also requires qualifying certain quality parameters before it could be used as a fuel to manufacture cement. Presently, RDF presented for use by RDF manufacturing facilities in waste to energy and cement plants is a misnomer for this material as it does not have the desired quality or consistency in it for using it as a source of fuel.[20] While RDF shows great potential for contributing towards AFR mix, its use is greatly hampered because of lack of quality standards. This has adversely affected viability and thus the business models.

The RDF prepared from MSW may have different physical and chemical properties depending on where they are sourced, particularly with reference to their ash, chlorine, sulphur, and water contents.[21] As RDF is a material that has been prepared from MSW that has not been efficiently segregated at source, utilizing the same would hamper the kiln combustion process. Thermal substitution rates must be carefully determined also considering levels of sulphur and chlorine in raw materials. Unregulated input of alternate fuels with these volatiles would lead to preheater blockages necessitating a gas bypass system[22] and high moisture content in fuels could limit kiln capacity. This necessitates that RDF needs to be pre-processed to make it fit to be used in a cement kiln. Table 2[23] identifies a broad guideline defining parameters of raw MSW and RDF acceptable by cement plants. These could be further refined and tuned to cater to individual cement plants through detailed discussions with RDF manufacturer.

---

[18]Report by Global Alliance for Incinerators Alternatives.

[19]TERI-Suez Report on feasibility on co-processing in cement plants.

[20]Recommended Quality Standards for SRF and RDF from MSW to improve resource recovery in cement kilns —UlhasParlikar, Dy. Head, Geocycle India, ACC Limited.

[21]Hajinezhad et al. [4].

[22]White paper on Technical Guidelines on Environmentally Sound Pre-Processing Facilities To Prepare Homogenous Waste Mixes Suitable for Co-processing in Cement Kilns—IIP Network.

[23]Recommended Quality Standards for SRF and RDF from MSW to improve resource recovery in cement kilns —UlhasParlikar, Dy. Head, Geocycle India, ACC Limited.

Table 2 Broad guideline
defining parameters of raw
MSW and RDF

Parameter	MSW	RDF 1	RDF 2
Size	<400 mm	<75 mm	<35 mm
GCV	>1500 Cal/gm	2500–3500 Cal/gm	2500–3500 Cal/gm
Moisture	No limit	<25%	<25%
Ash	No limit	<20%	<20%
Chlorine	No limit	<1%	<1%

5 Effective Availability of Waste for RDF Co-processing

Most of the mega cities in India that generate in excess of 3000 tonnes may automatically qualify as attractive options for generating non-recyclable organic waste that could be processed into RDF and co-processed in cement plants. However, there are no cement kilns around these cities and even if RDF is produced the most likely user would be waste to energy plants.[24] An assessment of was performed to calculate the potential levels of TSR achievable by utilization of waste available within 100 km radius of cement plants.

5.1 Results of the Analysis

The outcome of the assessment suggests that by utilization of MSW within the radius of 100 km from cement plants, the Indian cement sector can achieve a TSR of 0.48%. This further suggests that area greater than 100 km radius may be explored for increasing substitution rates. The assessment further suggests that substitution of coal with RDF prepared by utilization of waste within 100 km radius of cement plants may yield a carbon reduction of 205,778 tCO_2/year. The observations of the assessment are tabulated below.

---

[24]TERI-SUEZ Report on Feasibility of co-processing RDF in cement plants.

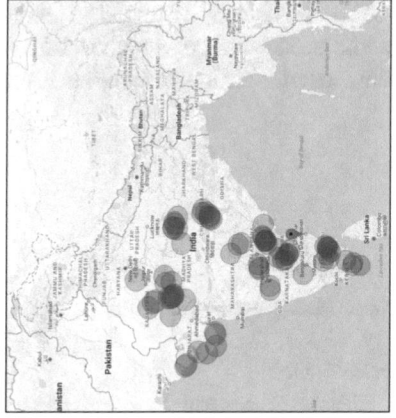

S.No	State	Area	Quantity (TPD)	Waste Available TPA	Available for RDF (20%)	RDF Yield (10%) TPA	Production of Cement
1.	Chhattisgarh	Raipur	450	3,86,900	77,380	7,738	9.79
		Bilaspur	130				
		Korba	130				
		Durg	130				
		Bhilai	220				
2.	Gujarat	Amreli	74	1,51,658	30,332	3,033	14.43
		Junagadh	150				
		Rananv	76				
		Rajkot					
		Gandhidham	86				
		Jamnagar	29.5				
3.	Karnataka	Gulbarga	263	15,69,135	3,13,827	31,383	9.57
		Belgaum	180				
		Bagalkot	159				
		Tumkur	110				
		koppal	78				
		Bijapur	109				
		Hubli	400				
		Bangalore	3000				
		Satna	62.5				
		Ratlam	50				
4.	Madhya Pradesh	Jabalpur	344	2,57,321	51,464	5,146	20.54
		Sagar	64				
		Katni	69.49				
		Singrauli	45				
		Rewa	70				
5	Rajasthan	Jaipur	1100	10,99,380	2,19,876	21,988	33.97
		Kota	510				
		Udaipur	231				
		Baran					
		Ajmer	131				
		Beawar	110				
		Jodhpur	550				
		Nagaur	120				
		pali	110				
		Alwar	110				
		Jhunjhunu					
		Bhilwara	150				
6	Tamil Nadu	Thanjavur	288	11,14,345	2,22,869	22,287	20.97
		Tiruchirappalli	405				
		Dindigul	100				
		Madurai	680				
		Erode	250				
		Salem	350				
		Thoothukudi	217				
		Tirunalvei	150				
		Nagerkoil	105				
		Kurichi					
		Thiruppur	508				
7	Telangana & AP	Anantpur	125	23,50,600	4,70,120	47,012	29.75
		Kadapa	206				
		Naigonda	60				
		Guntur	450				
		Kurnool	210				
		Hyderabad	4000				
		Warangal	200				
		Karimnagar	125				
		Ramagundam	165				
		Khammam	120				
		Rajahmundry	280				
		Anantpur	125				
		Vijaywada	374				

% Cement Production	Major Cement Producing States	Production of Cement(Million Tonnes Per annum)	Amount of Clinker (Ratio=1.35) (Million Tonnes	RDF available (10%) Tonnes	Energy from RDF (Kcal)	Energy converted in MTOE	TSR	Energy required for Cement production (Kcal)	Energy converted in MTOE
5.44	Chattisgarh	9.79	7.25	7,738	19345000000	0.001932566	0.38%	5.0763E+12	0.507122
8.02	Gujarat	14.43	10.69	3,486	8714375000	0.0008670566	0.12%	7.4822E+12	0.747474
5.32	Karnataka	9.57	7.09	31,383	78456750000	0.007837829	1.58%	4.9622E+12	0.495726
11.42	MP	20.54	15.21	5,146	12866067500	0.00128532	0.12%	1.065E+13	1.063972
18.89	Rajasthan	33.97	25.16	21,988	54969000000	0.005491403	0.31%	1.7614E+13	1.759646
11.66	TN	20.97	15.53	22,287	55717250000	0.005566153	0.51%	1.0873E+13	1.086246
16.54	Telangana AP	29.75	22.04	47,012	1.1753E+11	0.011741247	0.76%	1.5426E+13	1.54105
Total	77.29		102.98	1,39,039	3.47598E+11	0.034725084	0.48%	7.2084E+13	7.201236

6   Lack of Informal Sector Integration

The informal waste management sector consists of individuals, families working in the waste management sector who are neither recognized or sponsored by the mainstream waste management authorities.[25] Largely, the sector makes its living by collecting recyclables from the waste stream and deriving economic value from it. The tag 'informal' is synchronous to lack of formal identity provided to them by the system.[26] Presently the formal system of waste collection and valorization doesn't involve informal waste pickers. The waste collected by informal waste pickers is sorted by them and sold to itinerary waste buyers for selling it for further recycling. While sorting the waste, the most soiled kind of recyclables are not picked up by waste pickers or discarded. Proper integration of informal sector in formal waste collection mechanism would not only protect their livelihood but also improve waste collection by RDF processors.

7   Lack of Workable Business Models

The potential that co-processing technology has displayed requires a concerted effort by all responsible stakeholders for realization of on-ground success. The technology has been proven in many countries to be one of the efficient methods for waste management and mode for fossil fuel substitution. A workable business model is a must to ensure long term sustainability of the mechanism. The Cement Manufacturers Association (CMA)[27] observes that, *to encourage the Industry to utilise more fly-ash and increasing AFR usage, they must be supplied 'Free-of-Cost' on the world-accepted principle 'Polluter Pays'*. The RDF manufacturing facilities feel that they may not be assured of a sale to a cement plant even if they establish a process line to cater superior cement grade RDF.[28] These are concerns raised by stakeholders taking into account individual concerns and could be addressed by devising a sound business model. The following technical models[29] present scenarios for implementation in the Indian context.

---

[25] Economic Aspects of the Informal Sector in Solid Waste Management—GIZ, 2010.

[26] Velis et al. [5].

[27] Cement Manufacturers Association—Annual Report 2013–14.

[28] Ramky enviro engineers—Enhanced Usage of Alternate Fuels and Raw Materials In Cement Industry—http://www.cmaindia.org/templates/magicaltie/images/conference/Technical-V/Mr-Varun-Dilip-boralkar.pdf.

[29] Adapted from financial viability and barriers for implementing AFR project—Ulhas Parlikar, Dy Head, Geocycle India.

8 Model 1: Direct Pre-processing Model

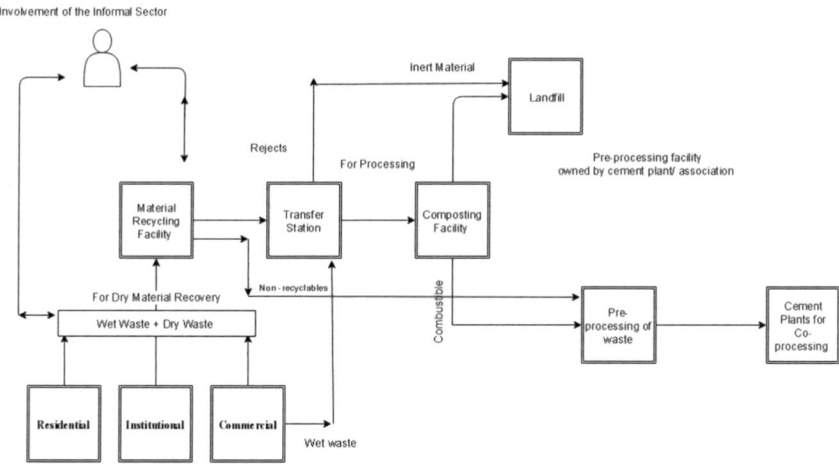

Direct pre-processing model is implemented in cases where RDF plants are yet to be setup near cement plants/clusters. The model also ensures recycling of waste through the Material Recycling Facilities (MRF) as per the waste hierarchy. Combustible waste from composting facilities could be delivered at the gates of cement for pre and co-processing. The pre-processing facility may be installed as a part of the cement plant which could be used to treat waste before co-processing at the plant. The advantages of implementing such a model are that the government can save the investment on setting up RDF plants. This model will also ensure seamless use of RDF use in cement kilns with no quality concerns being raised by cement companies. However, a cost sharing/business model with a sound cash flow is to be devised for ensuring a profitable business for all entities including the municipality.

9   Model 2: Cluster Model

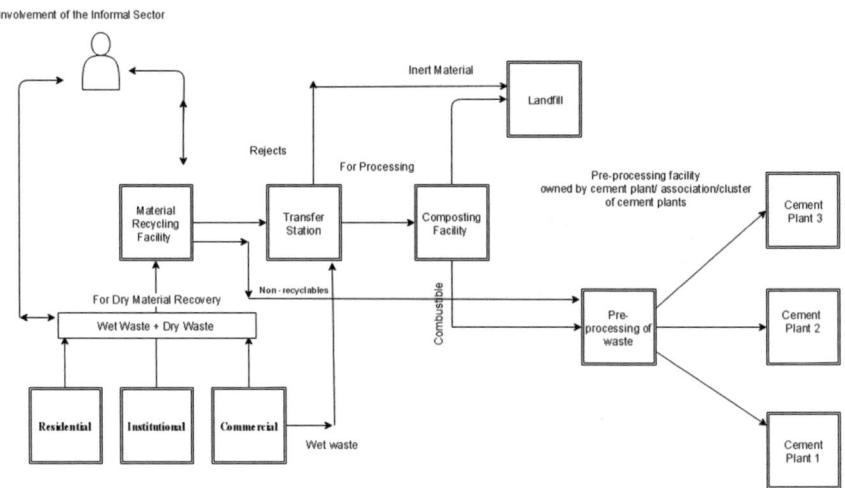

The cluster model could be utilized where multiple cement plants are in close proximity. The pre-processing plant could be owned by an independent body or an association of cement plants and the resultant RDF could be utilized by multiple cement plants in the vicinity. The advantage of such a model is that the cost of installing multiple pre-processing facilities could be saved. Further, the transportation costs to individual plants could be minimized.

10   Model 3: Market Mature Model

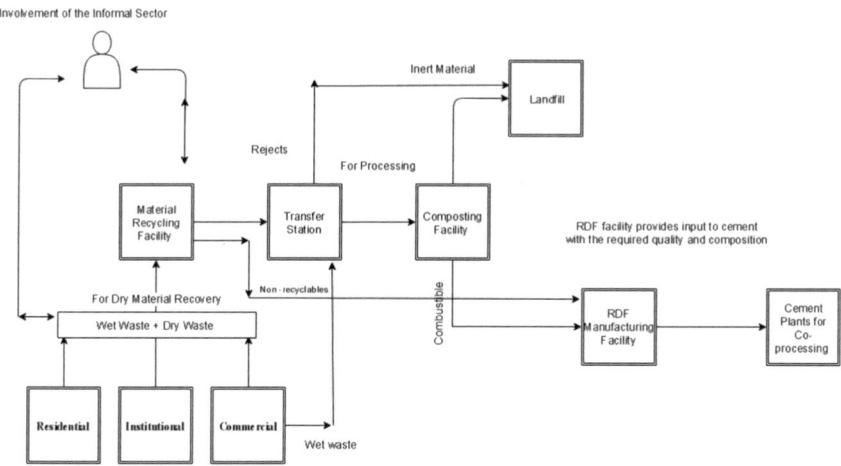

The market mature model attempts to tap the expertise and maturity of the RDF plant setup to cater to individual clients/cement plants in terms of the required quality. In this model, the RDF plant processes the technical capability of catering to the cement plant as per the required quality parameters. Advantage of this model is that the market for the RDF plant is wide open as it is able to cater to the quality requirements of various clients including the cement plants. However, at this stage, this may be a futuristic model, the RDF plants are to improve on their abilities to cater to the different clients based on their individual requirements.

11   Conclusion

RDF co-processing is a beneficial technology that could address a low carbon strategy for cement plants and efficient waste management for ULBs. However, it is key to understand the barriers affecting its replication potential and address the same. It is evident from the assessment that waste available within 100kms of cement plant would contribute to 0.48% TSR in the cement sector and the plants have to look beyond this area in order to increase levels of substitution. Quality could be a major deterrent in spelling levels of RDF use in plants. Cement plants/associations and ULBs must collectively look into devising innovative business models to ensure sustainability of co-processing.

References

1. Handbook of Technologies-Solid Waste Managmement. (2016). *Swachh Maharashtra Abhiyan* (1st ed.). https://swachh.maharashtra.gov.in/Site/Upload/GR/SWM_Handbook.pdf.
2. Boesch, M. E., Hellweg, S. Identifying improvement potentials in cement production with life cycle assessment.
3. Georgiopoulou, M., Lyberatos, G. (2017). Life cycle assessment of the use of alternative fuels in cement kilns: a case study. *Journal of Environmental Management*. http://dx.doi.org/10.1016/j.jenvman.2017.07.017.
4. Hajinezhad, A., Halimehjani, E. Z., Tahani, M. Utilization of refuse-derived fuel (RDF) from urban waste as an alternative fuel for cement factory: a case study.
5. Velis, C. A., Wilson, D. C., Rocca, O., Smith, S. R., Mavropoulos, A., Cheeseman, C. R. An analytical framework and tool ('InteRa') for integrating the informal recycling sector in waste and resource management systems in developing countries.

Economic Feasibility and Environmental Sustainability of a Community Scale Multi-component Bioenergy System

Dipam Patowary, Trinakshee Sarmah, Gunajit Dev Sarma, Bharat Terang, Rupam Patowary and Debendra Chandra Baruah

Abstract Sustainable development of rural sector has been a prime concern of the present World including India. Technology plays an important role for rural development. Appropriateness of any technology with reference to economy and environmental sustainability is a prerequisite. In this context, scope of a biogas-based bioenergy technology cluster is investigated with reference to a representative rural area in Assam. The bioenergy system is conceptualized for production of (i) fuel for domestic cooking to replace fuel wood, (ii) organic fertilizer for crop farming replacing chemical fertilizer and (iii) mushroom as saleable commodity. The relevant information pertaining to an on-going work is used to investigate the (i) economic feasibility, (ii) potential environmental benefits and (iii) potential benefits to generate rural employment. About 70 m^3 biogas (375 kg of fuel wood) from about 2 tonnes of cattle dung available in the rural cluster of 33 households would completely replace the fuel wood used for domestic cooking. Moreover, the solid part of digested slurry along with surplus crop residues available in the village would produce about 60 tonnes of organic fertilizer per annum which is equivalent to 16 tonnes of chemical fertilizer. Further, the use of about 150,000 L of surplus digested liquid slurry (separated from the solid part) for production of about 10 tonnes of mushroom appears potential to enhance the overall economy of the system. The benefits of all the individual components (viz., biogas as cooking fuel, solid digested slurry as organic fertilizer and liquid slurry as supplement for enhancing mushroom production) are analyzed in terms of economy and environmental sustainability considering the available information for the rural clusters considered in the study.

Keywords Sustainable development · Biogas plant · Mushroom cultivation · Economic feasibility

D. Patowary · T. Sarmah · G. D. Sarma · B. Terang · R. Patowary · D. C. Baruah (✉)
Energy Conservation Laboratory, Department of Energy, Tezpur University, Tezpur, Assam, India
e-mail: baruahd@tezu.ernet.in

© Springer Nature Singapore Pte Ltd. 2020
S. K. Ghosh (ed.), *Energy Recovery Processes from Wastes*,
https://doi.org/10.1007/978-981-32-9228-4_20

1 Introduction

Implementing clean and affordable cooking fuels is a critical social objective for improving the environment of rural areas in India. Generally, a traditional cook stove meets the cooking requirements of an average rural household. On the other hand, these cook stoves result in indoor air pollution which in turn causes health problems and in severe cases, death. Women make up a sizeable portion of the sufferers of primitive indoor cooking practices as they spend a lot of time in drudgery and cooking [1–3].

The universal access to clean cooking fuel in rural areas is not possible, especially in developing countries like India. However, the substitution of traditional sources of energy like firewood, chips, kerosene, coke, coal etc. with renewable sources of energy like biogas is still desired. In rural India, firewood and chips are principal sources of energy for cooking by more than three-quarters (76.3%) of households, LPG by 11.5%, and dung cake by 6.3%. About 1.6% of rural households do not have any arrangement for cooking. The remaining households use other sources of energy like kerosene (0.8%) and coke (0.8%) [4–6].

Swachh Bharat Abhiyan, launched in India on 2nd October, 2014 was aimed at creating clean villages by eliminating open defecation and handling solid and liquid wastes. In view of this, the Ministry of Drinking Water and Sanitation launched the GOBAR (Galvanizing Organic Bi-Agro Resources) scheme on 30th April, 2018 to convert the solid wastes from livestock into useful matter like cooking fuels (biogas and bio-CNG) and compost fertilizer. The various thrust areas of the GOBAR DHAN scheme include: proper handling of wastes, developing organic fertilizer to supplement the chemical fertilizers, improving the indoor air quality, generating bioenergy as well as generating employment [7].

In view of these thrust areas and also the problems of threat of exhaustion of natural resources such as firewood from forests and many cardio vascular diseases arising from inhaling the harmful pollutants from burning firewood indoors, an alternate solution has been discussed in this paper to tackle these issues. It has been envisaged that proper utilization of solid wastes like cow dung can solve the dual issues of solid waste management and cooking fuels [6, 8].

Utilization of solid wastes to produce biogas (for cooking) could be an effective way to tackle the dual issues of waste management and cooking. Surplus availability of cow dung in rural areas of India is a suitable feedstock option for biogas production [9]. Further, the byproducts (i.e., digested slurry) from biogas plant have sufficient fertilizer value [10–12]. Organic fertilizer like Vermicompost production from dried digested slurry is also reported to be beneficial considering the value addition of biogas digested slurry [13–15]. In addition, liquid portion of digested slurry is also reported to be an effective supplement for enhancement of mushroom production [16–18].

The present work is aimed at investigating the feasibility of installing a biogas-based bioenergy system in a typical rural area of Assam. Cow dung generated in the village will be estimated for providing sufficient biogas supply to the village community. The scope of utilization of digested slurry left behind from the system will be further investigated for testing economic feasibility of the system. Apart from this, environment sustainability of the system will also be addressed in addition to possible ventures of employment generation from the system for the villagers.

2 Materials and Methods

2.1 Description of the Village

The area considered for the present study is Jhawani village, located at a distance of 17 km from the Tezpur town of Sonitpur District, Assam. Jhawani village is under Bihaguri Development Block, located in the southern side of the Sonitpur district. A location profile of the village is presented in Table 1 and Fig. 1.

Table 1 Location profile of Bihaguri development block and Jhawani village

Location and area		Bihaguri block	Jhawani village
Location	Longitude	92°30′54.719″E and 92°43′55.025″E	26°37′43.637″N and 26°38′25.756″N
	Latitude	26°42′28.29″N and 26°36′46.709″N	92°41′39.35″E and 92°40′44.277″E
Area	(km^2)	155.43	2.09

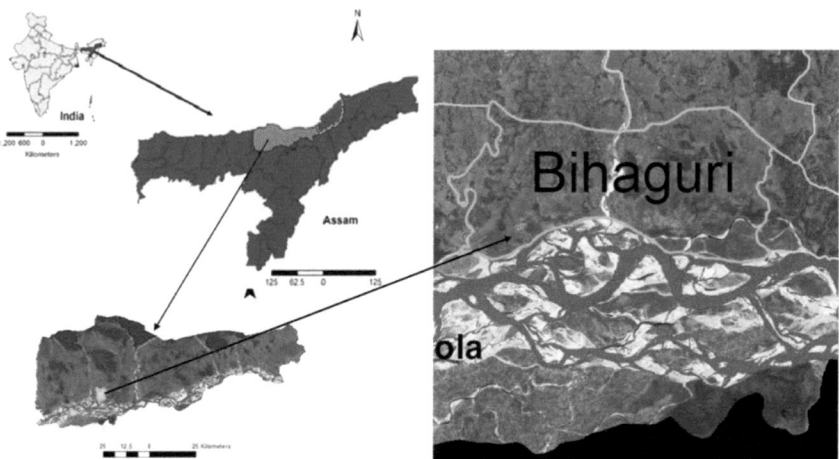

Fig. 1 Map showing Bihaguri development block

Jhawani has 33 households with a total population of 137 people. It is mainly an agrarian-based society with 89% of the population depending on farming for their livelihood. The remaining population mostly engage themselves in rearing cattle.

2.2 Cooking Energy Consumption

Firewood and liquid petroleum gas (LPG) are the two predominantly used energy sources in cooking. Firewood is available within the village, mostly within the homestead. *Sesbania javanica*, (a biomass cultivated in the village) grown almost in all household is mostly used as firewood for cooking and space heating. LPG is used by six households, cylinders being collected from *Bihaguri* (about 5 km from Jhawani village) facing difficulty in collecting and transporting. The estimated highlight of the cooking scenario in Jhawani is presented in Table 2.

2.3 Biomass Resource Availability in Jhawani

Jhawani village is an agriculturally dominated village with 85% of households depending on agriculture. The estimated amounts of available crop residue biomass mostly available for each household in the village are presented in Table 3. Some of the residues have competitive uses. However, about 50% rice straw, black gram straw, mustard straw is available as this amount is left over in the crop field. In addition, cattle rearing is also one major occupation in the village with 28 out of 33 households have cows. About 2 tons of cow dung is available in the village on the daily basis.

Table 2 Domestic energy consumption pattern in Jhawani village

Activity and energy source		Number of household	Population	Village total (MJ/day[a])	Daily consumption (kg/day)
Cooking fuel	Firewood	33	137	5970	375
	LPG	6	28	65.46	1.42

[a]Calorific values of firewood and LPG are considered as 16 and 46 MJ/kg

Table 3 Major crop residue biomass and cow dung availability

Crop residues/Cow dung	Months	Amount available (kg)
Rice straw	Oct (IV week)–Nov (III week)	20,768
Mustard straw	Jan (IV week)–Feb (III week)	11,281
Black gram straw	Jan (I week)–Feb (I week)	2274
Cow dung	Throughout the year	2204

Cow dung produced is generally used as a fertilizer for crops and the excess cow dung is sold in Jhawani village. Despite this, the handling of cow dung and disposal of the agro wastes is a problem to the households. Mostly the farmers are tempted to dispose the excess cow dung into dumping areas and burn the surplus agro residues which become a source of soil and air pollution.

2.4 Energy Recovery from Available Biomass

As discussed, the cooking need of the Jhawani village is mostly met by firewood. Despite inefficient cooking using firewood, collection and its preparation for fuel involves time and effort and most importantly indoor air pollution while cooking. The end result is the affected health of the women and children of the families. In order to tackle these issues; an effective solution is a necessity. Jhawani, being a resourceful village, utilization of available biomass in an environmentally effective manner can be sought as an option to replace the current cooking practice.

Prior to the consideration of an alternative fuel being generated through an environmentally sustainable system, the energy demand for cooking needs to be determined based on the time required for cooking which in turn depends on the size of the family. This basic information of the time required for cooking on a daily basis for each household are obtained from household surveys and information for the cooking demand has been estimated thereafter. Moreover, investigation on the availability of resource and selection of appropriate technology for conversion of these resources in an environmental and economical manner to meet the cooking demand of the village is a necessary step for feasible application and acceptance of the technology.

2.5 Description of the System

Considering the Jhawani village, a biogas-based bioenergy system can be a suitable technology that holds the potential to benefit the villagers economically as well as environmentally. A majority of Jhawani village's population is dependent of fuelwood for cooking accounting its consumption of up to 375 kg per day. A biogas system utilizing the local resources (Cow dung) generated in the village can act as a viable option in fulfilling the cooking requirement of the village. Moreover, fulfilling cooking needs of the village is not the only aspect of the biogas plant. Utilization of byproduct of the system into organic fertilizer and sellable value added products is itself an added benefit of such kind of bioenergy systems. Figure 2 shows the various components of a proposed multi-utility biogas-based bioenergy system considering a sample rural area like Jhawani village. The components of the system are explained in the following points.

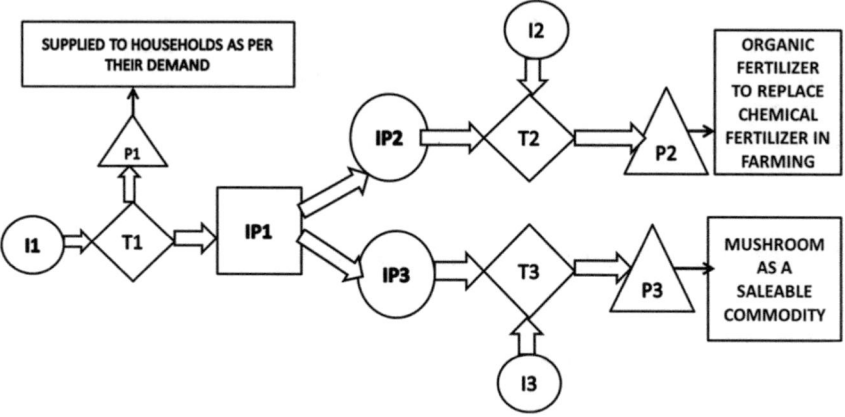

Fig. 2 Components of the multi-utility biogas-based bioenergy system

1. **Biogas plant (T1)**: An already proven and widely accepted technology (KVIC model) for biogas production from cow dung available in the Jhawani village can be set up in meeting the cooking energy requirement. Since, the system is expected to be a community scaled one, demand for cooking and assessment of cow dung generation in the village is prerequisite before deciding its size. Moreover, a community scaled biogas system comes with added auxiliaries for its daily operation. The amount of biogas production required to not only fulfill the cooking demand but also having a scope to power the auxiliaries for an uninterrupted supply of biogas needs to be done prior to the designing of the system. The sub components for such a biogas system includes:

   a. **Mixing Tank**: Cow dung (I1) and water is mixed using a mechanical mixer in the mixing tank. This process of mixing requires electricity supply.
   b. **Buffer storage vessel and Gas distribution**: Biogas generated (P1) in the floating dome is transferred to the pressure regulation buffer tank which pressurizes the gas using water column. Pressurized gas is then distributed to households through underground pipelines
   c. **Digested slurry handling system:** Digested Slurry (IP1)is separated into solid (IP2) and liquid portion (IP3) in a dewatering screw press machine. The size of the dewatering screw machine is considered based on the amount of daily available digested slurry from the biogas digester.
   d. **Power Generation system**: A power generator using biogas as fuel to supply power to motor (mixing cow dung and water), compressor (compression of biogas in buffer storage) and dewatering screw press.

2. **Organic fertilizer production unit (T2)**: The by-product of the biogas plant, i.e., digested slurry (digestate) can be used as a value added product by the production of organic fertilizer like vermicompost (P2) [13, 14]. Digested slurry is firstly separated into solid and liquid portion using a dewatering press. The

solid portion of the digested slurry (IP2) is then mixed with agro residues and vermi worms (I2) and it takes a period of 45–50 days to produce vermicompost. It has been reported that 30% of product can be recovered as vermicompost per unit kg of biomass (dried digested slurry, agro residues). The cost calculations have been taken from the already reported literature [15]. Vermicompost has an added advantage over chemical fertilizers due to the enhanced productivity of crops on application of it [15]. For an agriculturally dominated village like Jhawani, application of vermicompost over chemical fertilizers will not only enhance productivity of crops but also bring environmental sustainability.

3. **Mushroom production unit (T3)**: Liquid portion of the digested slurry (IP3) after being separated using a dewatering press can be used as a supplement for mushroom (P3) cultivation. It has been reported that the use of liquid digested slurry enhances mushroom production (*Species: P. Florida*) as compared to conventional process of oyster mushroom production. Addition of liquid digested slurry to substrates, i.e., I3 (sterilized rice straws with mushroom spawns)is done for a period of 30 days after the spawning period of 25–30 days for enhanced mushroom production [17]. It has been reported that *P. florida* gave about 2.4 kg of mushroom per kg of substrate on application of liquid portion of digested slurry [16, 17]. Considering the multi-utility of the bioenergy system, production of mushroom as a saleable commodity in Jhawani village can not only improve the economic feasibility of the system but also generate employment for the villagers.

2.6 Economics of Multi-utility Biogas-Based Bioenergy System

The analysis of cost economics of the entire biogas-based system involving biogas utilization and activities associated with its byproduct utilization is necessary to test the feasibility of such system in any area. In the present study, cost economics is investigated considering the full utilization of biogas for cooking and powering system components as well as 100% utilization of digested slurry in the production of organic fertilizer and mushroom.

To estimate the cost, capital investment of a community biogas plant (CBP) with its size based on available cow dung and biogas demand, organic fertilizer production unit and mushroom production unit based on digested slurry utilization from the CBP has been considered.

The economic analysis of such a CBP has been carried out taking into consideration of two sub cost parameters, i.e., recurring and non-recurring cost. Specific key components to develop a production unit are identified under each cost parameter. The various components of non-recurring cost are (i) Capital investment of CBP (ii) Capital cost of organic fertilizer unit and mushroom production unit. While the various components of recurring cost are (i) chemicals/spawn//bags/tray/vermiworms/gunnybags/spades/nets/tools and accessories (ii) interest on investment (iii) insurance (iv) depreciation (v) manpower requirement in man days.

3 Results and Discussion

3.1 Energy Demand for Cooking and Estimation of Available Energy

Cooking time depends on the size of the family and it varies from two to five hours/day. Utilization of biogas as a cooking fuel can save the daily firewood usage up to 375 kg thereby reducing drudgery and time in fuel preparation for cooking. The daily availability of cow dung from all the households in Jhawani is estimated to be around 2204 kg. It has been estimated that there is a demand of 54.8 m^3 of biogas in meeting the cooking energy requirement of the village against a total potential of 88.16 m^3 from the available cow ding. The household-level biogas generation potential (cow dung as feedstock), demand for cooking and powering system components are shown in Table 4.

However, considering the demand, 70 m^3 of biogas is sufficient in meeting the cooking needs and providing power to system components. This can be easily available if 80% of available cow dung is utilized. Therefore, a 70 m^3 CBP can be considered to be a suitable option for the selected Jhawani village.

3.2 Mass Flow in the CBP

A daily requirement of 1750 kg of cow dung is required for the production of 70 m^3 of biogas per day. The mass flow involved in the CBP is shown in Fig. 3.

Digested slurry (Digestate) available from the CBP is expected to be separated into liquid and solid portion which accounts for a daily availability of 0.54 tonnes of solid and 1.2 tonnes of liquid digested slurry. Apart from biogas for cooking to the households, utilization of the solid and liquid portion of the digested slurry can produce about 60 TPA of organic fertilizer and 10 TPA of mushroom, respectively.

3.3 Economic Feasibility of the Biogas-Based Bioenergy System

The investigation of the economic feasibility of the aforementioned conversions of cow dung into biogas and digestate in value added products needs to be investigated.

Table 4 Household-level biogas generation potential and demand for cooking

Household (nos.)	Cattle dung availability (kg/day)	Biogas potential (m³/day) (@0.04 m³/kg of cattle dung)	No of family members	Biogas demand for cooking (m³/day) (@0.45 m³/capita/day)	Biogas for cooking and powering system components + 10% excess (m³)	Firewood savings (kg/day)
33	2204	88.16	137	54.8	70	375

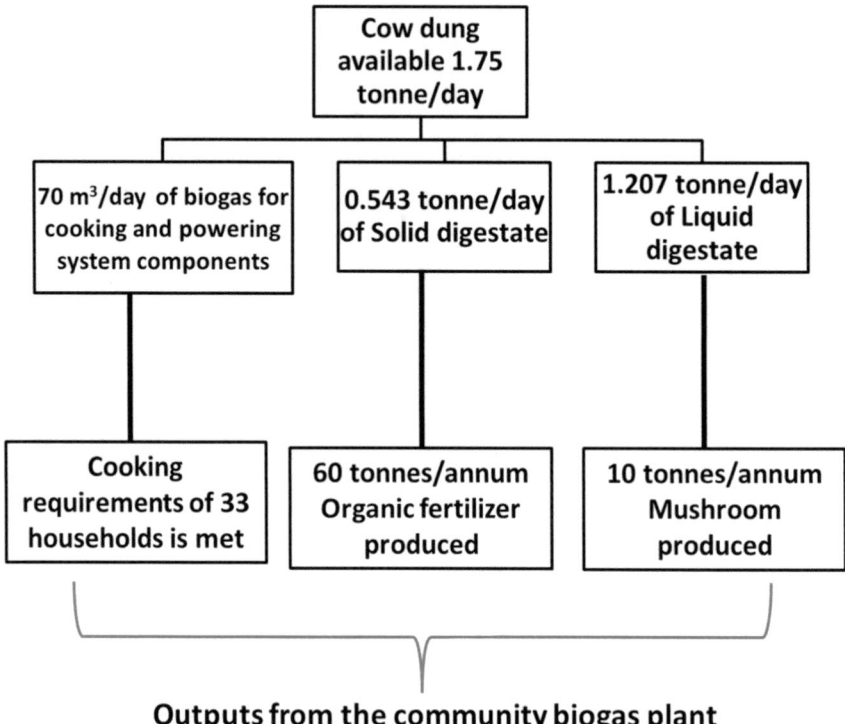

Fig. 3 Mass flow diagram of CBP

Utilization of cooking using biogas can save up to 375 kg of daily firewood requirement. With all the time invested in collection and preparation of fuel, cooking using firewood is not a very dependable source of cooking energy. Moreover, it is also a source of indoor air pollution. Considering these aspects, biogas from the community biogas plant supplied to the families at a rate as low as INR 200/month can be an agreeable alternative considering the reliability of the technology and overall benefits over firewood cooking.

The economic analysis of CBP has been carried out taking into consideration two sub cost parameters, i.e., recurring and non-recurring cost. Specific key components to develop a production unit are identified under each cost parameter. The various components of non-recurring cost are capital investment of CBP and mechanical screw press for separation of the digestate. The various components of recurring cost are cost of cow dung, manual labor for cow dung preparation and cow dung feeding (INR 60,000 PA @ 200 man days) and maintenance cost of CBP. The revenue generated from the biogas is INR 79,200 per annum. Capital investment of a 70 m^3 community biogas plant with provision of gas supply to households is considered in consultation with the reported literature [19].

Revenue generated by selling the biogas produced from the community biogas plant alone cannot form a cost effective venture. The solid digestate and the liquid digestate are also required to for an effective economic model of the CBP.

3.3.1 Organic Fertilizer from Solid Digestate

Based on the potential production of organic fertilizer of 60 TPA, a total of five harvests will be required from an organic fertilizer production unit of 12 tonne capacity. The production of organic fertilizer requires decomposable crop residues, solid digestate, vermi worms and soil. Crop residues, as already mentioned are available in surplus quantities in the village.

The non-recurring cost is the capital investment of the unit. The recurring cost involves the cost of crop residues already available in the village, manual labor for vermi bed preparation and water spraying (INR 60,000 PA @ 200 man days) and packaging organic fertilizer (INR 9000 @ 30 man days), maintenance cost of the production unit, depreciation. Two laborers are employed for the production and packaging of the mushrooms and one laborer is employed for cow dung preparation and cow dung feeding in the biogas plant.

The revenue obtained from the 60TPA organic fertilizer unit is expected to be INR2.97 lakhs. The cost calculations have been referred from the already reported literature [15].

3.3.2 Mushroom Enterprise

For the present study, 2 kg of mushroom production is considered per unit kg of substrate. Crop residue like rice straw is required to be sterilized prior to the spawning process. As already reported, there is a surplus availability of rice straw in the village.

Liquid digestate available from the CBP is sufficient for producing 10 TPA of mushroom. The prevalent market value of the mushroom is INR 100 per kg. In the mushroom enterprise, the recurring costs are manufacturing costs which include spawn, polybags, rice straw and packets for packaging, maintenance cost of mushroom, depreciation manual labor for production and packaging mushroom (INR 40,800 PA @ 136 man days). The cost calculations for the mushroom production unit have been referred from the already reported literature [18]. The detailed cost analysis considering all the avenues discussed is presented in Table 5.

From the table, it can be stated that the revenue of the CBP has attained an amount of INR 13.33 lakhs per annum. Considering multi-faceted horizons of a biogas system, a balanced economic model can be projected.

Table 5 Detailed cost analysis of the CBP

S. no.	Particulars	
Fixed cost		
1	Capital cost of biogas plant (Rs.)	2,500,000
2	Capital cost of vermi compost unit (Rs.)	140,000
3	Mechanical screw press for separation of digestate	500,000
4	Mushroom production unit	400,000
	Total fixed cost (Rs.)	3,540,000
	5% depreciation	177,000
	7% interest	247,800
	15% insurance and tax	531,000
	Total	4,495,800
Recurring cost of the plant		
5	Packaging of vermicompost in gunny bags (Rs.)	16,620
6	Manual labor for packaging (Rs.)	9000
7	Manual labor for vermi bed preparation, bed spraying (Rs.)	60,000
8	Cost of cow dung (Rs.)	219,000
9	Mushroom production cost (spawn, polybags, rice straw + packets for packaging)	94,000
10	Manual labor for production and packaging mushroom	192,000
11	Manual labor for cow dung preparation and cow dung feeding (Rs.)	96,000
12	Maintenance cost of cbp (Rs.)	90,000
13	Maintenance cost of vcu and mushroom (Rs.)	16,200
	Total recurring cost (Rs.)	792,820
Revenue		
	Biogas (Rs.)	79,200
	Vermicompost (Rs.)	254,580
	Mushroom (Rs.)	1,000,000
	Total revenue	1,333,780

3.4 Environmental Sustainability of the CBP

The problems of threat of exhaustion of natural resources such as firewood from forests and many cardio vascular diseases arising from inhaling the harmful pollutants from burning firewood indoors have necessitated the substitution of sources of energy like firewood to cleaner sources of energy like biogas. Application of biogas will not only improve the cooking practice but also will have direct benefits like reduction of drudgery associated with fuel collection, better health of families. In addition, production of 60 TPA of organic fertilizer can save up to 2–2.7 TPA of Urea, 5–7 TPA of diammonium phosphate (DAP) and 7.2 TPA of muriate of potash

(MOP) [20]. It will not only eradicate the dependency on chemical fertilizers but also the productivity of crops will be enhanced. Moreover, entrepreneurial avenue like mushroom production has the potential of effective utilization of liquid digestate. The whole system of multi-utility features of biogas-based bioenergy can be considered as an environmentally sustainable model based on the benefits already stated.

3.5 Employment Generation of the CBP

Introduction of a bioenergy system with multiple avenues brings a plethora of man-power requirement to fulfill the tasks associated with the smooth operation of the system. Engagement of local people from the community for biogas generation, organic fertilizer and mushroom production will not only generate income for them but also the economic status of the working families will be improved.

4 Conclusion

The main aim of current study is to reduce the dependency of the people on the conventional sources of traditional energy like firewood for imparting a profound livelihood option for them. Feasibility of setting up a biogas-based bioenergy system is investigated in the current study with scope of employment generation from the system. The multi-faceted benefits availing from system viz., biogas for cooking, organic fertilizer production for enhancing crop productivity and mushroom production as an income generating venture have proved to be both economically and environmentally sustainable. A 70 m^3 community biogas plant has the potential of supply biogas supply to 33 households in a typical rural area like Jhawani village with an added option of 60 TPA of organic fertilizer production and 10 TPA of mushroom production. The biogas-based system has a payback period of six years.

References

1. Aayog, N. I.T.I. *Draft national energy policy*. Government of India, Version as on 27.06.2017.
2. Goal 7: Affordable and Clean Energy. In *Sustainable development goals*. United Nations Development Programme, http://www.undp.org/content/undp/en/home/sustainable-development-goals/goal-7-affordable-and-clean-energy.html. Accessed on September 15 2018.
3. Goal 8: Decent work and economic growth. In *Sustainable development goals*. United Nations Development Programme, http://www.undp.org/content/undp/en/home/sustainable-development-goals/goal-8-decent-work-and-economic-growth.html. Accessed on September 15 2018.

4. National Sample Survey Office. *Energy sources of Indian households for cooking and lighting*, NSS 66th Round (July 2009–June 2010).
5. Solid Waste Management in Rural Areas A Step-by-Step Guide for Gram Panchayats, http://www.nird.org.in/nird_docs/sb/doc5.pdf. Accessed on September 15 2018.
6. Sustainable energy for all, Global Tracking Framework, 2017, http://gtf.esmap.org/data/files/download-documents/eegp1701_gtf_full_report_for_web_0516.pdf. Accessed on September 15 2018.
7. Guidelines for GOBAR-DHAN (Galvanizing Organic Bio-Agro Resources Dhan) under Swachh Bharat Mission (GRAMIN), Ministry of Drinking Water and Sanitation, Government of India 2018.
8. WHO. (2016, February). *Fact sheet No. 292: Indoor air pollution and health.*
9. Rao, P. V., Baral, S. S., Dey, R., & Mutnuri, S. (2010). Biogas generation potential by anaerobic digestion for sustainable energy development in India. *Renewable and Sustainable Energy Reviews, 14*(7), 2086–2094.
10. Liu, W. K., Yang, Q. C., Du, L. F., Cheng, R. F., & Zhou, W. L. (2011). Nutrient supplementation increased growth and nitrate concentration of lettuce cultivated hydroponically with biogas slurry. *Acta Agriculturae Scandinavica, Section B-Soil & Plant Science, 61*(5), 391–394.
11. Liu, W. K., Yang, Q. C., & Du, L. (2009). Soilless cultivation for high-quality vegetables with biogas manure in China: Feasibility and benefit analysis. *Renewable Agriculture and Food Systems, 24*(4), 300–307.
12. Tambone, F., Scaglia, B., D'Imporzano, G., Schievano, A., Orzi, V., Salati, S., et al. (2010). Assessing amendment and fertilizing properties of digestates from anaerobic digestion through a comparative study with digested sludge and compost. *Chemosphere, 81*(5), 577–583.
13. Sørensen, P., Møller, H. B. (2009). Fate of nitrogen in pig and cattle slurries applied to the soil-crop system. In: F. Adani, A. Schievano, & G. Boccasile (Eds.), *Anaerobic digestion: Opportunities for agriculture and environment* (pp. 27–37). DiProVe University of Milan, Milan.
14. Adhikary, S. (2012). Vermicompost, the story of organic gold: A review. *Agricultural Sciences, 3*(7), 905.
15. Model scheme for vermi-composting units under agri-clinics, Department of Agriculture Cooperation & Farmers Welfare, http://agricoop.nic.in/sites/default/files/Vermicompost%20Production%20Unit.pdf Accessed on September 18 2018.
16. Malayil, S., Chanakya, H. N., & Ashwath, R. (2016). Biogas digester liquid—A nutrient supplement for mushroom cultivation. *Environmental Nanotechnology, Monitoring & Management, 6*, 24–31.
17. Ashwath, R., Chanakya, H. N., & Malayil, S. (2016). Utilization of biogas digester liquid for higher mushroom yeilds. *Procedia Environmental Sciences, 35*, 781–784.
18. Kataki, S., Sarma, G. D., Patowary, D., & Baruah, D. C. (2019). Prospects of utilization of liquid fraction of biogas digestate as substrate supplement for mushroom cultivation. In *Advances in Waste Management* (pp. 445–465). Springer, Singapore.
19. Nasery, V. (2011). *Biogas for rural communities*. Center for Technology Alternatives for Rural Areas, Indian Institute of Technology Bombay. Available at: www.cse.iitb.ac.in/~sohoni/pastTDSL/BiogasOptions.pdf. Accessed May 31 2012.
20. Kataki, S. (2017). *Assessment of bioenergy By-products (Anaerobic digestate and Biochar) as potential crop production inputs*. Doctoral Thesis.

Printed in the United States
By Bookmasters